CELL BIOLOGY RESEARCH PROGRESS

A CLOSER LOOK AT MEMBRANE PROTEINS

CELL BIOLOGY RESEARCH PROGRESS

Additional books and e-books in this series can be found
on Nova's website under the Series tab.

CELL BIOLOGY RESEARCH PROGRESS

A CLOSER LOOK AT MEMBRANE PROTEINS

TRISTAN B. MØLLER
EDITOR

Copyright © 2020 by Nova Science Publishers, Inc.

All rights reserved. No part of this book may be reproduced, stored in a retrieval system or transmitted in any form or by any means: electronic, electrostatic, magnetic, tape, mechanical photocopying, recording or otherwise without the written permission of the Publisher.

We have partnered with Copyright Clearance Center to make it easy for you to obtain permissions to reuse content from this publication. Simply navigate to this publication's page on Nova's website and locate the "Get Permission" button below the title description. This button is linked directly to the title's permission page on copyright.com. Alternatively, you can visit copyright.com and search by title, ISBN, or ISSN.

For further questions about using the service on copyright.com, please contact:
Copyright Clearance Center
Phone: +1-(978) 750-8400　　　　　Fax: +1-(978) 750-4470　　　　　E-mail: info@copyright.com

NOTICE TO THE READER

The Publisher has taken reasonable care in the preparation of this book, but makes no expressed or implied warranty of any kind and assumes no responsibility for any errors or omissions. No liability is assumed for incidental or consequential damages in connection with or arising out of information contained in this book. The Publisher shall not be liable for any special, consequential, or exemplary damages resulting, in whole or in part, from the readers' use of, or reliance upon, this material. Any parts of this book based on government reports are so indicated and copyright is claimed for those parts to the extent applicable to compilations of such works.

Independent verification should be sought for any data, advice or recommendations contained in this book. In addition, no responsibility is assumed by the Publisher for any injury and/or damage to persons or property arising from any methods, products, instructions, ideas or otherwise contained in this publication.

This publication is designed to provide accurate and authoritative information with regard to the subject matter covered herein. It is sold with the clear understanding that the Publisher is not engaged in rendering legal or any other professional services. If legal or any other expert assistance is required, the services of a competent person should be sought. FROM A DECLARATION OF PARTICIPANTS JOINTLY ADOPTED BY A COMMITTEE OF THE AMERICAN BAR ASSOCIATION AND A COMMITTEE OF PUBLISHERS.

Additional color graphics may be available in the e-book version of this book.

Library of Congress Cataloging-in-Publication Data

Names: Møller, Tristan B., editor.
Title: A closer look at membrane proteins / Tristan B. Møller, editor.
Description: New York : Nova Science Publishers, [2020] | Series: Cell
　biology research progress | Includes bibliographical references and
　index. |
Identifiers: LCCN 2020035223 (print) | LCCN 2020035224 (ebook) | ISBN
　9781536181494 (paperback) | ISBN 9781536185386 (adobe pdf)
Subjects: LCSH: Membrane proteins.
Classification: LCC QP552.M44 C56 2020 (print) | LCC QP552.M44 (ebook) |
　DDC 572/.696--dc23
LC record available at https://lccn.loc.gov/2020035223
LC ebook record available at https://lccn.loc.gov/2020035224

Published by Nova Science Publishers, Inc. † New York

Contents

Preface		vii
Chapter 1	Bitter–Sweet Story of the IGF Receptors in Cell (Mal)Functioning *Dragana Robajac, Miloš Šunderić, Nikola Gligorijević and Olgica Nedić*	1
Chapter 2	Simulating Membrane Proteins *N. K. Roy*	91
Chapter 3	The Commandments of Studying Integral Membrane Proteins *Raymond J. Turner*	121
Index		147

PREFACE

A Closer Look at Membrane Proteins opens with a description of the insulin-like growth factor system, with focus on the insulin-like growth factor receptors and functions associated with them. The data on membrane proteins, their N–glycome and oxidation status id related to the authors' findings on the receptors in different physiological and pathological conditions, such as normal and abnormal tissue growth and development.

Next, a review of the current methods used to prepare and study membrane proteins is presented, with focus on large scale simulations and special emphasis on scalable parallel methods.

In closing, commandments surrounding integral membrane protein expression and purification, integral membrane protein biochemistry, integral membrane protein functionality studies and integral membrane protein high-resolution structures are described.

Chapter 1 - The insulin–like growth factor (IGF) system contains multiple members including growth factors, their binding proteins and receptors. After binding of growth factors to their receptors, a cascade of signals is activated initiating a number of mitogenic and metabolic pathways. Being at the crossroad of different and sometimes opposing functions, dependent on structural modifications as well as cell surrounding, the IGF system represents an intriguing field of investigation.

It is involved in cell growth, proliferation and energy metabolism, but also in cell apoptosis. The IGF system will be described in the following chapter, with the focus on the IGF receptors and functions associated with them. The data on membrane proteins, their N–glycome and oxidation status will be related to the authors' findings on the receptors in different physiological and pathological conditions, such as normal and abnormal tissue growth and development. Placental and colorectal tissues will be used as examples.

Chapter 2 - Membrane proteins may contain a significant portion of their mass within the interior of the membrane or are only associated to the membrane surface. The transmembrane (TM) part can be helical or have a sheet topology. The TM part can have a variety of sizes, molecular weights and conformations. Membrane proteins govern biological processes such as energy conversion, transport, signal recognition and transduction. Up to 30% of the encoded proteins in the genome of all organisms are such proteins, and are 60% of all drug targets. Currently less than 1% of the protein structures deposited in the RCSB Protein Data Bank are membrane proteins. Simulating membrane proteins correctly provides the best way to study them. Ion channels are a class of membrane proteins where the passive transport is influenced by the membrane potentials. Many such ion channels have the selectivity filter and gating mechanism embedded in the membrane core. Asymmetric ion concentrations across the membrane also affect transport and protein functions. These are difficult to study. Large scale molecular dynamics (MD) with coarse graining in both the membrane lipid bilayer and in parts of membrane protein itself is generally the method used in any simulation study to understand the mechanisms of such proteins' structure-function relationships and dynamic modes. Principal component analysis of the protein is also frequently used. This chapter gives an overview of the current methods used to prepare and study membrane proteins. The focus is on large scale simulations with special emphasis on scalable parallel methods. Correctly relating molecular structures to the physiological properties of the protein is a major challenge in the field. All the effects of the inhomogeneous lipid bilayer, potentials, ion/anion concentrations, that cover both spatial and temporal scales must be included. This has challenges when systems have thousands of explicit atoms and require simulations on the micro-second scale. The authors review these challenges and explain methods that have been used to overcome the short comings of explicit MD simulations.

Chapter 3 - As a budding Biochemist, the author was introduced to Arthur Kornberg's ten commandments of enzymology. After 25 years of working in the field of integral membrane protein (IMP) structure-function, the author's trainees and students of their classes noticed that the author would make statements on IMP research in the form of a commandment, in the guises of these early Biochemistry commandments. Here the author shares my commandments around IMP expression and purification, IMP biochemistry, IMP functionality studies, and IMP high-resolution structures.

In: A Closer Look at Membrane Proteins
Editor: Tristan B. Møller
ISBN: 978-1-53618-149-4
© 2020 Nova Science Publishers, Inc.

Chapter 1

BITTER–SWEET STORY OF THE IGF RECEPTORS IN CELL (MAL)FUNCTIONING

Dragana Robajac[*]*, Miloš Šunderić,*
Nikola Gligorijević and Olgica Nedić
Institute for the Application of Nuclear Energy (INEP),
University of Belgrade, Belgrade, Serbia

ABSTRACT

The insulin–like growth factor (IGF) system contains multiple members including growth factors, their binding proteins and receptors. After binding of growth factors to their receptors, a cascade of signals is activated initiating a number of mitogenic and metabolic pathways. Being at the crossroad of different and sometimes opposing functions, dependent on structural modifications as well as cell surrounding, the IGF system represents an intriguing field of investigation. It is involved in cell growth, proliferation and energy metabolism, but also in cell apoptosis. The IGF system will be described in the following chapter, with the focus on the IGF receptors and functions associated with them. The data on membrane proteins, their N–glycome and oxidation status will be related

---
[*] Corresponding Author's E-mail: draganar@inep.co.rs.

to our findings on the receptors in different physiological and pathological conditions, such as normal and abnormal tissue growth and development. Placental and colorectal tissues will be used as examples.

Keywords: insulin–like growth factors, membrane receptors, placenta, diabetes, colon cancer

INTRODUCTION

The insulin–like growth factor (IGF) system consists of peptides (IGF–I and IGF–II), binding proteins (IGFBP–1–6), receptors (IGF–1R and IGF–2R) and IGFBP proteases, as presented in Figure 1. Being closely related to IGF peptides and IGF–1R, and due to high degree of homology and crossreactivity, insulin and insulin receptor (IR) can also be observed as the part of the IGF system, as will be discussed in the following sections.

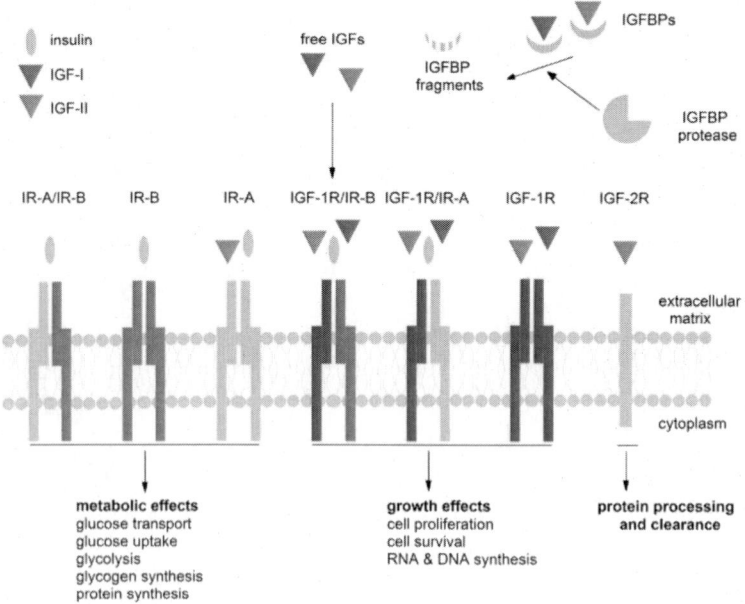

Figure 1. Members of the IGF system and the processes they control.

Peptides of the IGF System

Peptides of the IGF system are 7.5 kDa proteins, with amino acids arranged in four domains (B, C, A and D), and are mainly produced in the liver (Le Roith et al., 2001). IGF–I and IGF–II share 70% similarity, while the difference between IGFs and insulin is around 50% (Rinderknecht and Humbel, 1978). The main difference between IGFs and (pro)insulin is the existence of domains C and D (both are omitted from insulin while proinsulin contains domain C). Second difference originates from the specific amino acids in the IGFs located at positions 3, 4, 15 and 16, that are responsible for recognition and binding to IGFBPs. At last, difference between IGFs and insulin is derived from the existence of IGFBPs, which bind IGFs leaving rather minute amounts of free peptides, whereas binding proteins do not control insulin availability, allowing this peptide to be able to circulate freely (Annunziata et al., 2011).

IGF–I gene is located in the 12q region, and after birth its expression (and the production of IGF–I responsible for growth and proliferation) is under the control of growth hormone. *IGF–II* gene is located in the 11p region, whose anomalies result in abnormal foetal and postnatal growth, as well as increased risk of embryonic tumours in early life (Brioude et al., 2018; Wakeling et al., 2017). The term genomic imprinting is used to describe the monoallelic expression of genes based on their parental origin (Cassidy and Charalambous, 2018). Genomic imprinting is found to be of high importance as imprinting disorders share one common characteristic – developmental anomaly. As a result, altered growth and nutrient uptake are expressed already in early life. Placenta is a source of many proteins whose expression results in a genomic imprinting process, where some of these genes are found to be involved in the placentation process (Graves and Ranfree, 2013). *IGF–II* is maternally imprinted, and is normally inherited from father, whereas the loss of *IGF–II* imprinting is related to an increased cell proliferation and tumour risk (Livingstone, 2013; Kaneda et al., 2007). IGFs are expressed ubiquitously in all tissues where they act in an autocrine/paracrine manner, affecting cell growth, differentiation and proliferation, mitogenesis and metabolism. Additionally, the expression of other members of the IGF system (e.g., IGFBPs, IGFBP proteases and IGF–Rs) regulates the bioavailability of these peptides and hence their numerous functions.

IGF Binding Proteins

There are six proteins belonging to the family of IGF binding proteins (IGFBP–1–6) and there is also a family of IGFBP–related proteins, that share some structural similarities with IGFBPs but bind IGFs with much lower affinity (Brahmkhatri et al., 2015).

IGFBPs consist of 216–289 amino acids distributed in three domains (N, C, L), and it is the L domain that is not preserved between different IGFBPs, while other two share high degree of homology among IGFBPs and are responsible for high affinity binding of IGFs (Hwa et al., 1999). IGFBP–1, –3, and –4 bind both IGF peptides with the same affinity, while the affinity of IGFBP–2, –5 and –6 is higher for IGF–II compared to IGF–I (Firth and Baxter, 2002). IGFBPs are susceptible to different modifications: (i) IGFBP–1, –3 and –5 are phosphorylated (Jones et al., 1993; Hoeck and Mukku, 1994; Graham et al., 2007), (ii) IGFBP–3 and –4 are N–glycosylated (Firth and Baxter, 1999; Zhou et al., 2003), while (iii) IGFBP–5 and –6 are O–glycosylated (Graham et al., 2007; Neumann et al., 1998).

IGF peptides have higher affinity for IGFBPs than for receptors. Hence, IGFBPs regulate the availability of peptides acting as a storage system and also serve as protective molecules as they prolong the half–life of IGFs. Though all IGFBPs form binary complexes with IGFs, some IGFBPs (e.g., IGFBP –3 and –5) also form ternary complexes. The most abundant IGFBP in the circulation is IGFBP–3, which forms ternary complexes with IGFs and the acid labile subunit (ALS). After dissociation of the 150 kDa ternary complex, IGFs form binary complexes with other binding proteins, which further transport IGFs to target tissues (Le Roith et al., 2001).

Different (patho)physiological conditions may affect post–translational modifications of IGFBPs. For example, during pregnancy, placental alkaline phosphatase, produced by the syncytiotrophoblast, dephosphorylates blood IGFBP–1, thus reducing its affinity for IGF–I (Westwood et al., 1994), resulting in the prevalence of IGFBP–1 forms with lower affinities for IGFs, further leading to higher availability of free IGF–I (Forbes and Westwood, 2008).

RECEPTORS OF THE IGF SYSTEM

As already mentioned, due to high degree of homology between IGFs and insulin, receptors belonging to the IGF system are not only IGF type 1 (IGF–1R) and type 2 receptor (IGF–2R), but also insulin receptor (IR) and a hybrid receptor (HyR). Their structure, function and activating cascades will be discussed in the following sections and subsections, with their particular impact on gestation and cancer development.

Type 1 Insulin–Like Growth Factor Receptor – IGF–1R

The IGF–1R is a 420 kDa heterotetramer, a product of *IGF–1R* gene located in the 15q region. Furin cleaves IGF–1–pro–receptor into α (135 kDa) and β (95 kDa) subunits, while two pairs of αβ heterodimers are linked via disulphide bonds (Czech and Massague, 1982; Ward et al., 2001). IGF–1R is a transmembrane protein with two α–subunits located outside of the membrane, forming extracellular ligand–binding domain, while tyrosine kinase domains are located at the intracellular part of the protein formed by two β–subunits. Tyrosine kinase domains are located between juxtamembrane region and C–terminal, containing binding sites for phosphotyrosine from the signal molecules (Gatenby et al., 2013). It was shown that a heterozygous missense mutation in the *IGF–1R* gene can lead to the severe foetal growth restriction (IUGR) (Walenkamp et al., 2006), as can heterozygous mutations within the IGF–1R kinase domain

(Kruis et al., 2010) or extracellular second fibronectin III domain (Wallborn et al., 2010). As high circulating levels of IGFs are found in these children, their condition is reflecting reduced IGF–1R tyrosine phosphorylation or altered cell surface expression of IGF–1R.

IGF–1R contains 16 asparagine (Asn) residues that are potential N–glycosylation sites; eleven are located in α–subunit and five in β–subunit (Ullrich et al., 1986). They are involved in the processing, stabilisation and localisation of the receptor (Itkonen and Mills, 2013), while the Asn^{913} N–glycan is found to be responsible for the IGF–1R transport to the plasma membrane (Kim et al., 2012). The importance of N–glycans is also noted at the level of sialic acid as its absence on the IGF–1R (desyalilation of the receptor) impairs cell proliferation in response to insulin (Arabkhari et al., 2010), as nicely reviewed by Ferreira et al. (2018). Under regular physiological conditions, actions of the IGFs are mediated mainly via IGF–1R.

IGF–1R Cascades

Activation of the IGF–1R signalling pathway is a consequence of ligand–binding that induces conformational changes, facilitating autophosphorylation in the activation loop of the IGF–1R located in the α–subunit, leading to transphosphorylation of the opposing β–subunits (Hubbard and Till, 2000). As a result, activated IGF–1R further activates PI3K–AKT cascade and MAPK–extracellular signal–regulated kinase (MEK)–extracellular signal–regulated kinase (ERK1/2) cascade (Coolican et al., 1997; Duan et al., 2000; Imai and Clemmons, 1999).

Once activated, IGF–1R interacts with adaptor proteins including Src homology collagen (SHC) proteins (p46/p52/p66) and insulin receptor substrate (IRS) (Dupont and LeRoith, 2001). SHC is primarily involved in the activation of p21ras–MAPK, which plays important role in transduction of mitogenic signals initiated by different receptor tyrosine kinases, such as IGF–1R. After phosphorylation of IGF–1R, the receptor binds and phosphorylates SHC, leading to binding of adaptor protein growth factor–bound protein 2 (Grb2), which complexes with SOS, a p21ras guanine nucleotide exchange factor (Sasaoka et al., 2001). This action ends with ERK-1 and ERK activation, which leads to phosphorylation of cytoplasmic substrates and translocation and activation of transcription factors, enabling pro-proliferative and anti-apoptotic effects (Kolch, 2000). IGF-1R substrates are numerous: IRS, SHC, 14.3.3, CRK, CSK, PI3-kinase, SHP-2 phosphatase, Grb10 and others (Girnita et al., 2014).

IGF-1R can also utilize the components of the G-protein coupled receptor (GPCR) signalling, found to be essential for migratory and pro-survival functions controlled by IGF-I (Girnita et al., 2014). IGF-1R levels and functions are regulated by multiple post-translational modifications such as ubiquitination, sumoylation, phosphorylation, dephosphorylation (Girnita et al., 2014). Interestingly, IGF-1R propagates some completely opposing processes – growth and proliferation on one side and differentiation on the other, as well as cell adhesion on one side and motility on the other. Which process will be activated depends on the surrounding of the receptor and cell context. For example, IRS-1 is the main mediator of the mitogenic signals; however, cells that do not express IRS-1 will instead activate SHC and lead to differentiation (Romano, 2003).

Insulin Receptor – IR

The IR is also a 420 kDa tetrameric transmembrane tyrosine kinase, encoded from *IR* gene located in 19p region. It is composed of two α– and two β–subunits. Approximately 95% of IR can be found in this heterotetrameric form, out of which 75% resides in the plasma membrane (Hwang and Frost, 1999). Due to alternative splicing, two IR isoforms exist: IR–A (a receptor isoform missing exon 11 – $Ex11^-$, which encodes 12 amino acids from the IRs' ectodomain at the C–terminus of α–subunit) and IR–B (a receptor isoform with preserved exon 11 – $Ex11^+$) (Seino and Bell, 1989). IR–A isoform is preferentially expressed in foetal and cancer tissues, although it can also be found in the majority of other tissues (Frasca et al., 1999). IR–B is predominantly expressed during adult life, in an insulin targeted tissues, such as liver, muscle, adipose tissue and kidney. IR–A and IR–B differ in the tyrosine kinase activity and the degree of the IR's internalisation, signal transduction and distribution in the plasma membrane depending on the membrane content of cholesterol and caveolin (Uhles et al, 2003).

IR has 18 potential N–glycosylation sites (14 on α– and 4 on β– subunit) out of which 16 are regularly glycosylated (Sparrow et al., 2008). The presence of sialic acid is important for IR activation after insulin binding (Dridi et al., 2013), and changes in the sialic acid content can affect receptors' function (Pshezhetsky and Ashmarina, 2013).

IR Cascades

Similary to IGF–1R activation, ligand binding to IR results in the receptor autophosphorylation on cytoplasmic tyrosine residues and the phosphorylation of tyrosines of IRS proteins. Phosphotyrosine sites of IRS enable binding of the lipid kinase PI3K, responsible for the synthesis of phosphatidylinositol (3,4,5)–trisphosphate (PIP3) at the plasma membrane. This further activates phosphoinositide–dependent kinase that phosphorylates threonine residue of AKT, while the serine residue of AKT is phosphorylated by mTORC2. Once activated AKT activates/ phosphorylates other downstream substrates as the forkhead family box O (FOXO) transcription factors: (i) the protein tuberous sclerosis 2 (TSC2), which permits activation of mTORC1 and its downstream targets ribosomal protein S6 kinase (S6K) and sterol regulatory element binding protein 1c (SREBP1c), (ii) glycogen synthase kinase 3β (GSK3β) and (ii) the RabGAP TBC1 domain family member 4 (TBC1D4). As a result of IR activation, metabolic processes are initiated as well as cell growth and differentiation. Alternative substrates of IR are Grb10, Grb14 and suppressor of cytokine signalling (SOCS), which block IRS binding.

As can be seen for both described receptor protein kinases, phosphorylation is a crucial event for the activation or inactivation of IGF– 1R and IR, e.g., tyrosine phosphorylation activates while serine/threonine phosphorylation inactivates IR and IRS proteins. These mechanisms of IR activation and inhibition of its signalling pathways are in detail and critically described in an exceptional review of Haeusler and co–workers (2018).

IR–A expression is related to a decrease in metabolic signalling of insulin and the actions of IGFs and the signalling path they trigger upon activation of IR–A. Consequent to IGF–II and proinsulin binding to IR–A, mitogenic signals are activated resulting in cell growth, proliferation and survival, being important in foetal and cancer tissues. IR–B expression, on the other hand, is associated with an increase in metabolic signalling of insulin, being important during adult life. Importantly, both IR and IGF–1R can translocate into the nucleus, thus regulating biological responses at genomic level. This topic has exstensively been reviewed by Belfiore et al. (2017).

To increase the complexity of the IGF system, both IR and IGF–1R also have ligand–independent actions. For example, following ligand binding, due to catalytic activities of these receptors, anti–apoptotic signals are triggered enabling cell survival and resistance to apoptosis. However, unrelated to catalytic activities but only to receptors themselves, if lignads are absent and receptors are in a ligand–free form, they can support apoptosis. This effect can be reversed upon ligand binding (Belfiore et al., 2017).

IGF–1R/IR Hybrid Receptor – HyR

IR and IGF–1R share high degree of similarity, ranging from 40–95% depending on the domain (Lou et al., 2006). Accordingly, one IGF–1R αβ heterodimer can dimerize with one IR αβ heterodimer (either A or B isoform) forming a hybrid receptor – HyR (Benyoucef et al., 2007). HyR can be found in tissues rich in both IR and IGF–1R, and considering there are two forms of IR, there are also two forms of HyR. Although HyR consists of the halves of two closely related receptors, there are indications that HyR is functionally much closer to IGF–1R, irrelevant to splicing, as it has rather low affinity for insulin and readily binds IGFs (Benyoucef et al., 2007).

Type 2 Insulin–Like Growth Factor Receptor – IGF–2R

IGF–2R, also known as the cation–independent mannose–6–phosphate receptor, is a 270 kDa transmembrane protein encoded by the *IGF–2R* gene located in the 6q region. It mostly consists of the extracellular domain, while the transmembrane and cytoplasmic domains are much smaller (Lobel et al., 1988). IGF–2R is mostly expressed during foetal development (Sklar et al., 1992). As other members of the IGF system, IGF–2R is also subjected to post–translational modifications. Extracellular domain contains 19 potential N–glycosylation sites, out of which at least two are glycosylated (Lobel et al., 1987). IGF–2R serves as a clearing route for IGF–II as, when bound to IGF–2R, it is directed towards lysosomal degradation (Kornfeld, 1992). There are, however, some indications of the potential signalling pathway which involves IGF–2R.

IGF–2R Cascades

Unlike the IGF–1R and IR, IGF–2R does not contain tyrosine kinase activity or an autophosphorylation site. It is proposed that IGF–2R signalling is mediated by transactivating G protein–coupled sphingosine 1–phosphate receptors and that IGF–2R is involved in the ERK1/2 activation (El–Shewy et al., 2007). Although it is thought that IGF–2R serves only for IGF–2 clearence and degradation, there are speculations that IGF–II binding to IGF–2R may be involved in mediating mitogenic effects in term placental explants (Harris et al., 2011). Exact mechanism of the potential signalling pathway mediated by IGF–2R has yet to be elucidated.

Physiology

Mitosis, cell growth, differentiation, migration, transformation and apoptosis are processes controlled by the IGFs. The control is exerted during embryonal development, postnatal life, maturation from childhood until adult age and ageing. As mentioned, bioavailability and activity of IGFs is regulated by a network of IGFBPs, their proteases and IGF–Rs. In contrast to IGFs, insulin is mainly involved in metabolic processes (such as regulation of glucose concentration, protein and lipid metabolism), but mutual cross–reactivity enables IGFs to trigger metabolic responses and insulin to support cell growth (King et al., 1980). Malfunctioning in the IGF system may be associated with many pathophysiological states including cancer (Novosyadlyy and Le Roith, 2012; Nimptsch et al., 2019).

In contrast to insulin, secreted by pancreas, the majority of IGFs in the circulation is derived from the liver although many cells and organs can produce them locally. IGFs from the circulation exert predominantly endocrine functions, playing a role of a hormone. Peptides secreted locally exert paracrine and/or autocrine activities, expressing a role of cytokines or local growth factors. Liver is the principal organ of synthesis of IGFBP–1, IGFBP–2 and IGFBP–3, but considerable quantities are produced by other organs as well, whereas synthesis of IGFBP–4, IGFBP–5 and IGFBP–6 occurs in a number of organs (Blum et al., 2018; Clemmons, 2018). IGFBPs are associated with lipid and carbohydrate metabolism, development of atherosclerosis, bone and skeletal muscle metabolism (Clemmons, 2018). Common to all IGFBPs is their capacity to control the amount of free, biologically active IGFs and, thus, to limit their presentation to cell membrane receptors. IGFBPs also protect them against proteolysis and assist in their trafficking within an organism.

IGFBPs can associate in specific complexes with other proteins beside IGFs and with other biomolecules, such as those from an extracellular matrix and glycosaminoglycans (Russo et al., 1997; Liu et al., 2014). Some IGFBPs can bind directly to cell membranes via their receptors, activating pathways other than those activated by insulin and IGFs, although they can still carry IGF ligands (Ingermann et al., 2010; Clemmons, 2018). This activity was considered as IGF–independent until recently, but a caution was drawn since some of the interactions of IGFBPs influence IGF signalling within the same cell. IGFBPs can be transported in the nucleus where they interact with nuclear proteins causing alteration of cellular physiology (Bach, 2018).

The portion of free IGFs in healthy adult persons is not greater than 1% (Juul, 2003). In contrast to insulin, stored in pancreatic granules and released upon metabolic demand, IGFs are mostly stored in blood as IGF/IGFBP complexes. High affinity of IGFBPs for IGFs keeps the equilibrium between protein–bound and free form of these peptides. Upon demand for IGFs, IGFBP proteases modify IGFBPs reducing their affinity and enabling take–over by membrane receptors. Some IGFBPs are said to inhibit and the others to potentiate the activity of IGFs – the difference originates from the difference in the affinity of certain IGFBP compared to the affinity of receptors for IGFs (Le Roith, 2003; Clemmons, 2018).

It is worth mentioning that the IGF system is the only one having so many specific binding proteins (beside six high–affinity, there are several low–affinity), suggesting the need for very sensitive regulating mechanisms and fine tuning in respect to the actions of IGFs (Haywood et al., 2019). Regardless of the high degree of homology between insulin and IGFs, IGFBPs do not bind insulin (Annunziata et al., 2011). Physiological concentrations of insulin in healthy adults are in the range of pM whereas the concentrations of IGF–I and IGF–II are in nM (Juul, 2003; Heinemann, 2010). Thus, since IGFs are present in 100 to 1000–fold greater concentrations than insulin, their availability must be rigorously controlled.

IGF–I is a mediator of the activity of growth hormone (GH), as GH is an up–regulator of the *IGF–I* gene expression (Annunziata et al., 2011). In contrast to IGF–I, IGF–II is not regulated by GH (Kaplan and Cohen, 2007). Insulin/IGF signalling has been identified as one of the most important pathways that control the lifespan (Novosyadlyy and Le Roith, 2012). Blood levels of IGF–I are related to age (O'Connor et al., 1998), but they markedly differ between healthy individuals. Total IGF–I, IGFBP–3 and their ratio within one individual, however, show only small changes with age over time (Janssen, 2019). Although it is generally assumed that lower concentrations of IGF–I correlate with longevity, experimental findings do not consistently support this assumption. Lower protein intake during life may favour longevity through a process that regulates the concentration of circulating IGF–I. There may be a specific optimal age–dependent "set point" for each individual for the GH/IGF system which co–determines survival (Janssen, 2019). Enhanced signalling through the GH/IGF axis was noted to accompany an age–related disease – cancer

(Anisimov and Bartke, 2013). Therefore, maintaining equilibrium within the IGF system seems to be crucial for healthy living and ageing, and mechanisms that regulate lifespan and tumour incidence are mutually linked.

IGF–II was shown to exert growth–promoting actions in placenta and it influences prenatal growth and development of foetus (Cianfarani, 2012). A physiological role of this peptide, however still remains insufficiently known. Overexpression of the *IGF–II* gene can result in enlarged organs and the entire body size at birth (Kadakia and Josefson, 2016). Specific polymorphisms of *IGF–II* have been related to an increased weight, obesity, metabolic complications and hypertension (Gaunt et al., 2001; Gu et al., 2002; Faienza et al., 2010). A degree of *IGF–II* methylation at birth seems to be a crucial factor for development of such changes in early childhood (Perkins et al., 2012) and it was proposed to be considered as a biomarker of intrauterine programming (Cianfarani, 2012).

IGF–1R binds IGF–I with high affinity. It also binds IGF–II and insulin but with six and hundred fold lower affinity (Le Roith 2003; Annunziata et al., 2011). IR binds IGF–I with hundred fold lower affinity than insulin (Ullrich et al., 1986). A hybrid IR/IGF–1R has twenty times greater affinity for IGF–I than for insulin (Sakai et al., 2002). Tumour cells often demonstrate up–regulation or increased activity of IGF–1R (Novosyadlyy and Le Roith, 2012). As mentioned, IR exists in two isoforms where the form IR–A binds insulin and IGF–II, whereas IR–B binds predominantly insulin. The presence of specific isoforms may be associated with tumour development. Aggressive thyroid cancers, for example, overexpress IR–A, IGF–II and IGF–1R (Vella and Malaguarnera, 2018). IGF–2R binds IGF–II, but also mannose–6–phosphate (Man–6–P) and Man–6–P N–acetyl glucosamine (Olson et al., 2014; Nadimpalli and Amancha, 2010). IGF–II, after binding to IGF–2R, is most often degraded, so IGF–2R may be seen as a tumour–suppressor (Scott and Firth, 2004). Independently of the rest of the IGF system, IGF–2R acts as a lectin and regulates intracellular compartmentalisation of acid hydrolases containing Man–6–P residues (Olson et al., 2014; Nadimpalli and Amancha, 2010).

Post–translational modifications of IGFBPs and IGF–Rs, such as glycosylation, phosphorylation, oxidation and others can influence their affinity for IGFs. Phosphorylated IGFBP–1 has high affinity for IGF–I and usually inhibits its activity. During pregnancy, however, it is dephosphorylated by placental alkaline phosphatase to generate isoforms with lower affinity and consequently, increased IGF bioavailability (Solomon et al., 2014). In foetal blood, IGFBP–1 is the principal IGFBP. Maternal diabetes is associated with reduced IGFBP–1 phosphorylation in cord serum, suggesting that diabetes–related changes may additionally increase IGF–I bioavailability and stimulate foetal growth (Loukovaara et al., 2005). Phosphorylation of Ser^{111} enables IGFBP–3 to induce cell apoptosis (Jafari et al., 2018). Phosphorylation and O–glycosylation of IGFBP–5 affect its binding to heparin but not to IGFs (Graham et al., 2007). Glycosylation of IGFBP–3 does not seem to influence IGF binding or formation of protein complexes, but it influences the partitioning of IGFBP–3 between the extracellular milieu and the cell surface (Firth and Baxter, 1999). Post–translational modifications of the IGFBP–binding partners also affect the formation of protein complexes. Oxidation of fibrinogen, for example, reduces its interaction with IGFBP–1 in patients with diabetes mellitus, which may be important for the duration of bleeding and the speed of wound healing (Gligorijević et al., 2017). Ageing or colon cancer caused altered glycosylation of alpha–2–macroglobulin results in decreased binding of IGFBP–2, increasing the amount of free, physiologically active form of IGFBP–2 (Šunderić et al., 2019).

Both IR and IGF–1R belong to a tyrosine kinase family of receptors and they are therapeutic targets for the treatment of malignancy. Tumour cells develop resistance to targeted therapies over time by activating alternative signalling pathways. Enzymatic alteration and regulation of N–linked glycosylation process are seen as novel targets for developing approaches to sensitize tumour cells to cytotoxic therapies (Contessa et al., 2008; de–Freitas–Junior et al., 2017). Inhibition of N–linked glycosylation of receptors in patients with congenital disorders of glycosylation (CDG) was found to impair receptor processing and surface localization (Klaver et al., 2019). Reduced fucosylation of IGF–1R, due to impaired activity of fucosyltransferase, was shown to suppress proliferation, epithelial–mesenchymal transition, migration and invasion of specific placental cells (Yu et al., 2019). Pathophysiological conditions characterised by an increased oxidative stress, such as colorectal carcinoma, may induce oxidation of IGF receptors reducing their affinity for IGFs (Nedić et al., 2013).

Growing evidence suggests that IGFBP–1 and IGFBP–2 are favourably linked with insulin sensitivity and preclinical data implicate their direct involvement in the regulation of insulin signalling and adiposity (Haywood et al., 2019). These two IGFBPs have been under investigation as therapeutic targets for obesity, metabolic disorders and diabetes. Cancer treatment, on the other hand, most often includes strategies which enable lowering of the IGF concentration and inhibition of IGF–1R signalling (Ryan and Goss, 2008; Caban et al., 2019). A major future need is to identify cell membrane receptors and intracellular proteins which bind individual forms of IGFBPs and the signalling pathways that are activated following these interactions. This information will be useful in creating new therapeutic approaches for altering the activity of IGFs in different diseases.

PLACENTA

The main barrier as well as the bond between mother and the foetus is placenta, a multifunctional organ that represents the crucial determinant for the foetal growth and development. Placenta is the filter that enables transport of oxygen and all nutrients in the direction mother–foetus, but also a gland secreting hormones (estrogen, progesterone, growth hormone, and human placental lactogen). These hormones additionally support communication between mother and the foetus, in order to regulate maternal physiological adaptation to pregnancy and, more importantly, to fulfil different foetal needs. Placental characteristics, both morphological and functional (such as vascularisation, thickness, cell composition, presence and the abundance of different transporter/signalling molecules, hormones, etc.) alter throughout gestation (Fowden et al., 2009; Sandovici et al., 2012).

To understand the processes in the placental cell, the structure of placenta is given in Figure 2.

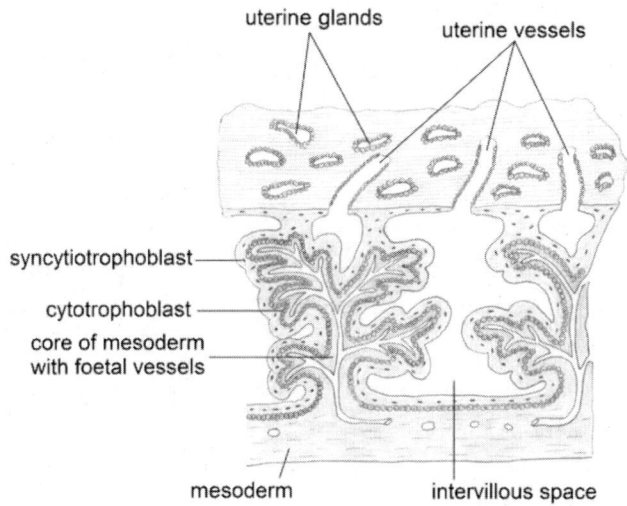

Figure 2. Schematic representation of human placental structures.

In chorionic villi, cytotrophoblast proliferates and differentiates into extravillous trophoblast or syncytiotrophoblast (Forbes and Westwood, 2008). Extravillous trophoblasts are oriented towards maternal endometrium into which they migrate and invade and remodel spiral arteries enabling the flow of oxygen and nutrients to the placenta and foetus. Syncytiotrophoblast is multinucleated layer that ensures nutrient and gas exchange and also serves as a protective barrier (Kingdom et al., 2000). Underneath the syncytiotrophoblast is the mesenchymal core that contains placental capillaries and different cells such as fibroblasts (Kingdom et al., 2000). As this multinucleated layer has no transcriptional activities, it is reconstituted due to continuous proliferation, differentiation and fusion of cytotrophoblasts (Forbes and Westwood, 2008). The rate of trophoblast turnover is associated with different tissue pathologies and is related to impaired foetal growth – both enhanced and reduced (Jansson and Powell, 2006). Placental nutrient transporters are localized on the syncytiotrophoblast, which is orientated towards mother and is soaked in the maternal blood (Forbes and Westwood, 2008).

The IGF System in Healthy Placenta

IGFs and IGFBPs

IGF–I is known to regulate differentiation of cytotrophoblasts to syncytiotrophoblasts and extravillous trophoblasts (Bhaumick et al., 1992; Milio et al., 1994; Lacey et al., 2002; Aplin et al., 2000). IGF–I that is synthesized in the villous mesenchyme stimulates extravillous trophoblast migration in a paracrine manner (Lacey et al., 2002). Placenta also produces IGF–II, which has important role in promoting trophoblast invasion (Hamilton et al., 1998) by inhibiting IGFBP–1 and TIMP3, molecules synthesized by the decidua to limit trophoblast infiltration into maternal tissues (Irwin et al., 2001). IGFBP–3 is the only IGFBP expressed by trophoblasts, fibroblasts of the villous stroma and the amnion and chorion laeve of the foetal membrane (Rogers et al., 1996; Han et al., 1996).

IGF–Rs and IR

IGF–1R is located in trophoblast, villous endothelium and mesenchymal core (Fang et al., 1997; Holmes et al., 1999). In the first trimester of pregnancy IGF–1R is more expressed in the proliferating cytotrophoblasts, whereas at term the main expression site of IGF–1R is the basal membrane of the syncytiotrophoblast and the villous cytotrophoblast – thus, the expression exerts stpatial–temporal change (Maruo et al., 1995; Hills et al., 2004). Being located at the basal membrane and the villous cytotrophoblast, IGF–1R is able to bind foetal IGF–I and IGF–II, whose levels can be elevated as a consequence of maternal diabetes. As a result, in such pregnancies, foetal IGF–I may overstimulate processes dependent of this peptide (Hiden et al., 2009).

IGF–1R and IR (as well as HyR) are abundantly expressed in placenta. In the beginning of gestation, in the first trimester, IR is predominantly expressed on the microvillous membrane of the syncytiotrophoblast, and to a lesser extent in the villous cytotrophoblasts, oriented to the maternal circulation. With gestation advancing, IR expression shifts towards the placental endothelium directed to the foetal blood. Consequently, it is maternal insulin that triggers placental IR in the first trimester of pregnancy, while throughout the pregnancy it is slowly being displaced with foetal insulin, shifting insulin–dependent processes from mother to foetus (Desoye et al., 1994; Hiden et al., 2006; Hiden et al., 2009). At term, IR localisation on the proliferation sites indicates its involvement in the vascular growth which is in accordance with the *in situ* data showing enhanced branching angiogenesis in gestational diabetes mellitus (Jirkovská et al., 2002). Elevated levels of foetal insulin can stimulate endothelial cells proliferation and vascular branching via binding to IR present on the sites of villus ramification (Hiden et al., 2009).

IGF–2R is expressed in the microvillous and plasma membranes of trophoblasts (Fang et al., 1997), and after proteolytic cleavage a soluble form of IGF–2R is released. Binding of soluble IGF–2R to IGF–II leads to degradation of this peptide and, hence, inhibition of its effects. It is reported that the molar ratio of IGF–II and soluble, circulating form of IGF–2R, are related to placental development and the weight of the newborn (Ong et al., 2000). In experiments performed using human trophoblast–derived choriocarcinoma cell line (BeWo) and placental explants, it was demonstrated that IGF–II protected IGF–2R–expressing BeWo cells from apoptosis, and the salvation of cytotrophoblast from apoptosis was found to be significant in a tissue with normal IGF–2R expression (Harris et al., 2011).

Exploring the N–glycosylation of placental membrane proteins, we demonstrated that gestation affects overall content of N–glycans, by increasing the abundance of core–fucosylated and multiantennary N–glycans meanwhile lowering the abundance of bisected biantennary N–glycans and terminal α2,3–sialylation (Robajac et al, 2014). During gestation (i) the content of total fucosylated, core–fucosylated biantennary

N–glycans and α2,6–sialylated N–glycans decreases in IR, (ii) the content of total fucosylated and α 2,6–sialylated N–glycans decreases, whereas the content of core–fucosylated biantennary N–glycans increases in IGF–1R, and (iii) overall fucosylation of IGF–2R increases together with the content of core–fucosylated biantennary N–glycans (Robajac et al., 2016a). The observed changes were not in accordance with the general pattern of glycosylation of placental membrane proteins, and even receptors differed between themselves.

Placental membrane proteins can be and are differently glycosylated. Most often these (glyco)proteins have been isolated using non–ionic (Triton or Tween) or anionic detergents (sodium dodecyl sulphate, SDS). However, considering that glycans affect polarity of (glyco)proteins, and other modifications contribute with their specific groups, it was demonstrated that the choice of a detergent may be crucial for the isolation of specific membrane (glyco)protein. Receptors of the IGF system are not an exception and one must bear this in mind when designing and conducting experiments that are based on the analysis of membrane proteins (Robajac et al., 2017).

IGF–Rs/IR Signalling in Placenta

It was demonstrated *in vitro* that IGF–I and IGF–II prevent apoptosis and enhance proliferation, migration and invasion of human placental villous explants, primary trophoblast cultures and trophoblast cell lines from the first trimester and term, as reviewed by Sferruzzi–Perri and co–workers (2017). Quantum dot experiment revealed the existence of a trans–syncytial pathway that allows mitotic signals to penetrate from maternal side to the inner progenitor cells (from syncytium into the cytoplasm of the underlying cytotrophoblast), which proliferate in order to assist placental and foetal growth (Karolczak–Bayatti et al., 2019). Proliferative effects of IGFs are mediated through IGF–1R activation and subsequent triggering of the MAPK signalling pathway, whereas anti–apoptotic effects are mediated via PI3K–AKT signalling pathway (Forbes et al., 2008). IGF–II effects on the first trimester human extravillous trophoblast invasiveness (stimulation of cell migration) are mediated via IGF–2R signalling, involving inhibitory G proteins and activation of the MAPK pathway (McKinnon et al., 2001).

Trophoblast migration and invasion are induced via IGFs binding to IGF–1R, and possibly to IR and, again, subsequent activation of the MAPK and PI3K–AKT signalling pathways (Mayama et al., 2013; Diaz et al., 2007). In the primary human trophoblasts, amino acid transporter activity, mediated by insulin and IGF–I, relies on the mTOR pathway (Roos et al., 2009). Insulin–related processes include the control of cell survival, differentiation and proliferation, as well as amino acid metabolism (Ruiz–Palacios et al., 2017). The glucose uptake is regulated by insulin only in the first trimester of pregnancy, while in the third trimester of pregnancy it is not (Ericsson et al., 2005). Insulin binding to IR in placenta activates Ras–ERK and the IRS–PI3K–AKT–mTOR pathways (Knofler et al., 2005; Colomiere et al., 2009). The PI3K–AKT–mTOR pathway affects cell apoptosis and proliferation, but its main responsability is for the nutrient metabolism. The role of AKT in placenta is unknown, although in peripheral tissues AKT triggers mechanisms for the translocation of glucose transporter (GLUT4) from cytoplasm to the membrane. This mechanism is still poorly understood, as GLUT1, and not GLUT4, is present in placenta (Ruiz–Palacios et al., 2017).

Preeclampsia and Intrauterine Growth Restriction

Preeclampsia (PE) is a condition characterised with increased blood pressure, proteinuria as well as other systemic disorders, diagnosed in women after 20th week of gestation, and is a leading cause of foetal and maternal mortality, with the incidence of 5% worldwide (Mol and Roberts, 2016). Insufficient invasion of trophoblasts and further incomplete remodelling of spiral arteries are regularly present in preeclamptic placenta. As a result, placental ischemia, low oxygen levels and oxidative stress are found in this type of placental pathology (Kanasaki and Kalluri, 2009). What is more, when complicated, PE can lead to IUGR, and additionally to acute kidney injury, thrombocytopenia, haemolysis and placental abruption, a condition called haemolysis–elevated liver enzymes and low platelet number, also known as HELLP syndrome (Wang et al., 2009).

Reduced foetal growth is closely related to aberrant placental development (Sibley et al., 2005). There are several reports on lower blood concentrations of IGF–I in mothers as well as lower expression of placental IGF–I (Peng et al., 2011; Kharb et al., 2016; Dubova et al., 2014; Kharb et al., 2017), and increased oxidation of placental IGF–1R (Robajac et al., 2015). Recent findings indicate that hypermethylation of IGF–I promoter is associated with PE (Ma et al., 2018). Levels of IGF–1R were also found to be decreased in reduced foetal growth (Laviola et al., 2005) and elevated in pregnancies complicated by macrosomia (Jiang et al., 2009). Compared to normal placentas, placentas complicated with IUGR are characterised by decreased content of IGF–1R, selective impairment of the IRS–2/PI3K pathway, and reduced p38 and c–Jun N–terminal kinase activation (Laviola et al., 2005). An increase in the placental protein content and an increase in the response to IGF–I of IGF–1R, IRS–1 and AKT was reported in small for gestational age placentas, which is, according to Iñiguez et al., a compensatory mechanism in response to IUGR (2014). The same authors recently reported that Klotho mRNA and protein concentration vary due to foetal growth and gestational age and down–regulate the activation induced by IGF–I on IGF–1R and AKT (Iñiguez et al., 2019).

PE is known as hypoxic condition, and it was reported that hypoxia decreases expression of PI3K–AKT and mTORC1 signalling in trophoblast cell lines (Yung et al., 2012). Low oxygen tension and IGF–I can maintain the multipotency and proliferation of placental mesenchymal stem cell via the IGF–1R signalling (Youssef et al., 2014). The same authors also demonstrated that culturing placental mesenchymal stem cells under low oxygen tension modifies IGF signalling through the IGF–1R or IR via ERK1/2 and AKT, also showing the involvement of these kinases in the regulation of stem cell destiny (Youssef and Han, 2016). They also indicated that multipotency of these cells can be mediated by IGF–I and IGF–II actions either via IGF–1R and/or IR (Youssef and Han, 2016). An increase in the content of IR and a decrease in the IGF–1R and IGF–2R content in placentas originating from mothers with diabetes (DM) and from those diagnosed with PE with IUGR was also reported (Robajac et al., 2015). Common to both DM and PE with IUGR is the oxidative stress, and elevated levels of protein carbonyls were reported for samples originating from either pathology (Robajac et al., 2015). Contrary to the general increase of carbonyls in placenta, isolated receptors were affected differently. As can be expected based on their high degree of homology, both IR and IGF–1R carbonylation was increased in PE complicated with IUGR, while DM had no effect on the carbonylation status of these receptors. IGF–2R carbonylation levels were not affected in either of the investigated pathologies suggesting greater resistance of this receptor to oxidative stress (Robajac et al., 2015). The glycan content of membrane proteins was also affected by these conditions (consequences of DM will be discussed in the next section). The overall fucosylation and the content of high–mannose N–glycans with more residues was decreased while the presence of paucimannosidic and high–mannose structures with lower number of mannose residues was increased in the placentas from pregnancies complicated by PE (Robajac et al., 2016b). Interestingly, changes observed at the level of receptors were found only in the case of IR α2,6–sialylation, which was decreased due to pathology (Robajac et al., 2016b). These findings undoubtedly demonstrated that there is no general pattern being followed by all proteins of the same origin (i.e., membrane proteins), and that each one of them is an entity that should be carefully characterised and its impact on/by different pathology investigated independently of other proteins. Knowing the role of the receptors of the IGF system, it is expected that the observed changes most probably affect downstream processes, as these conditions are characterised with newborns being labelled as either large (DM) or small (IUGR) for gestational age.

Novel findings also indicate a decrease in the expression of *IGF–2R* in placentas from pregnancies complicated by idiopathic IUGR (Harris et al., 2019). Placental expression of IGF–2R was found to be related to changes in the expression of homeobox genes that control cellular signalling pathways responsible for increased trophoblast cell apoptosis, one of the characteristics of IUGR (Harris et al., 2019). All these new reports on the IGF–2R are new arguments in favour of the receptors' role unrelated to simple IGF–II degradation.

DIABETES

Anomalies in insulin signalling pathway which may occur either due to the lack of insulin or a reduced sensitivity to insulin of its target tissues, lead to several glucose–related metabolic complications. Before clinical diagnosis of the type 1 diabetes, circulating autoantibodies against insulin, glutamic acid decarboxylase, the protein tyrosine phosphatase IA–2, and/or zinc transporter 8 can be detected. While individuals with a single autoantibody type frequently revert to negative status, reversion is rare in people with multiple autoantibodies. To diagnose type 1 DM, a positive result in at least two tests for autoantibodies is required. The presence of islet autoantibodies reflects immune B– and T–cell response to β–cell antigens. This autoimmune response to β–cells leads to the loss of β–cell mass and function giving rise to glucose intolerance and ultimately clinical symptoms of diabetes occur (Skyler et al., 2017). Current estimation is the existence of 425 million people with diabetes and another 352 million with impaired glucose tolerance worldwide, while the long–term prognosis for the 2045 is 629 million people with diabetes and 532 million of people with impaired glucose tolerance (IDF, 2017).

Type 2 DM develops when β–cells fail to secrete sufficient amounts of insulin in conditions of reduced insulin sensitivity. Although it has complex genetic and environmental aetiology, a major risk for development of type 2 DM is obesity. Ectopic fat deposition in liver and muscle leads to insulin resistance and insulin sensitivity in liver and muscle is improved by weight loss. Fat may also accumulate in the pancreas contributing to a reduced function of β–cells, inflammation and death of β–cells. Defects in insulin secretion are at least partially reversible with energy restriction and weight loss in pre–diabetes and recent–onset type 2 DM; however, in the case of long–standing diabetes, reversibility is very difficult (Skyler et al., 2017). While DM type 1 is a consequence of the immuno–mediated destruction of β–cells, and type 2 DM is mainly associated with an insulin secretory defects, there is evidence that mechanisms of these two types of diabetes overlap and have some common characteristics. For example, patients with type 2 DM also have reduced β–cell mass (Butler et al., 2003), and, in both types of diabetes, the stress response induced by hyperglycaemia may have a role in β–cell apoptosis (Laybutt et al., 2002). A reduction in the number of functional β–cells is very important for the development of hyperglycaemia and subsequent complications of diabetes. Hence, understanding the functional and differentiation state of β–cells is very important for defining subtypes of diabetes (Skyler et al., 2017).

IGF System in Diabetes

By binding to its receptor, insulin exerts multiple effects on metabolism, cell growth and differentiation, and studies have shown that components of the insulin signalling pathway also have a role in growth and secretion of pancreatic β–cells (Federici et al., 2001; Porzio et al., 1999). Some *in vitro* studies have shown that reduction of IGF–Rs enhances insulin sensitivity possibly by reducing the amount of HyR (Haywood et al., 2019). As already mentioned, insulin and IGF–I bind preferably to their specific receptors, although cross–reaction occurs since these hormones and their receptors are structurally very similar. Signalling pathways of these two hormones overlap at some points and may have the same consequences, such as translocation of GLUT4 transporter on the cell surface (Figure 3).

This overlapping mechanism can explain why in normal, healthy conditions IGF–I acts with insulin to lower glucose levels (Sjögren et al., 2001). During the development of insulin resistance, when glucose level is still within a normal range, there is an increase in the circulating IGF–I bioactivity. When a condition with impaired fasting glucose level develops, IGF–I bioactivity reaches a plateau. At the time when a person is finally diagnosed with type 2 diabetes, IGF–I bioactivity declines (Brugts et al., 2010). Some studies have shown that IGF–I levels are positively associated with insulin sensitivity and negatively with insulin secretion and diabetes in general (Teppala et al., 2010; Succurro et al., 2009). A large cross–sectional study in Denmark suggested a "U–shaped" association between IGF–I blood concentrations and insulin sensitivity, since both high and low IGF–I levels were associated with insulin resistance, compared to subjects with an intermediate IGF–I levels (Friedrich et al., 2012). There is, however, no firm biological explanation for this observation. It is suggested that the concentration of IGF–I alone is not sufficient for the assessment of this complex pathological condition. IGF–I is currently not considered as a potential therapeutic agent – while it lowers blood glucose levels it also shows some negative side effects (Moses et al., 1996).

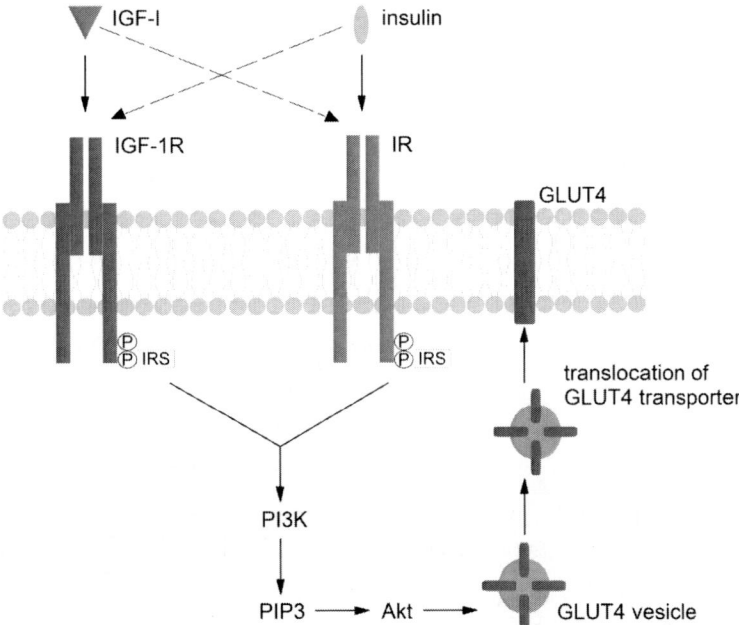

Figure 3. Involvement of the IR and IGF–1R in GLUT4 translocation.

A discovery that two insulin receptor isoforms have distinct functional characteristics created a hypothesis that alterations in the relative amount of these two isoforms may have a causal role in the development of insulin resistance, which is a feature of type 2 diabetes. Studies that explored the expression of two mRNA transcripts, for Ex11$^-$ and Ex11$^+$, offered opposing results. Three studies reported that the expression of these two mRNA species is changed in skeletal muscle of both pre–diabetic insulin–resistant and type 2 diabetic subjects when compared to persons with normal insulin sensitivity (Mosthaf et al., 1991; Mosthaf et al., 1993; Norgren et al., 1993), whereas other three studies failed to confirm these results (Benecke et al., 1992; Anderson et al., 1993; Hansen et al., 1993). Different results were also obtained when relative amounts of two isoforms were analysed. Two studies showed that the expression of the IR isoforms is changed in isolated adipocytes and in skeletal muscle membranes in subjects with type 2 DM (Sesti et al., 1991; Kellerer et al., 1993). A third study reported that the expression of the low–affinity IR–B form is significantly increased in fat and skeletal muscle from the obese and the type 2 diabetic subjects, when compared to non–obese controls (Sesti et al., 1995). An increased expression of the IR–B form significantly correlated with the body mass index and fasting glucose levels. However, there is also a report not confirming any difference in the relative proportion of the two IR isoforms in the skeletal muscle of healthy, obese persons without diabetes and obese persons with type 2 DM (Benecke et al., 1992). A caution must be taken, as small number of subjects involved in the last study might explain reported findings and the absence of difference between the investigated groups. In patients with insulinoma, the expression of the IR–B isoform is significantly increased in their skeletal muscles compared to healthy persons. This increase is positively correlated with the concentration of plasma insulin and negatively with insulin sensitivity (Sbraccia et al., 1996). Noted inconsistencies between reported studies may be a result of differences in PCR protocols, sites of biopsies, control groups used for comparison, metabolic status of the examined subjects and possible contamination of muscle tissue specimens.

IR Gene Mutations and Insulin Resistance

Insulin resistance syndromes originate from IR malfunction (Saito–Hakoda et al., 2018; Ardon et al., 2014). *IR* gene is located on the short arm of chromosome 19 and is composed of 22 exons. Mutations that affect insulin receptor coding gene lead to insulin–resistant syndromes such as Leprechaunism, also known as the Donohue syndrome, Rabson–Mendenhall syndrome and type A insulin resistance syndrome. The first two are autosomal recessive disorders, characterised by intrauterine and postnatal growth retardation, dysmorphic features, altered glucose homeostasis, and early mortality. Type A insulin resistance is autosomal dominant disorder and is characterised, besides insulin resistance, by *acanthosis nigricans* and hyperandrogenism such as polycystic ovarian syndrome. Since type A insulin resistance includes hyperandrogenism, this diagnosis is usually reserved for females, however, diagnosis could be generalized to men as well. Men can have insulin resistance with *acanthosis nigricans*, but without other clinical features such as lipoatrophy (Taylor et al., 1992).

The most severe of the insulin resistance syndromes is Leprechaunism. It is characterised by IUGR, loss of glucose homeostasis and hyperinsulinemia. Dysmorphic features are also present and include prominent eyes, thick lips, upturned nostrils, low–set posteriorly rotated ears, thick skin with lack of subcutaneous fat, distended abdomen, and enlarged genitalia in males and cystic ovaries in females (Longo et al., 1995). Cells from these patients have also markedly reduced insulin binding. Less severe syndrome is Rabson–Mendenhall. This condition was first described in three siblings with dental and skin anomalies, early dentition, coarse senile–looking faces, intellectual disability, prognathism, thick fingernails and *acanthosis nigricans*. Even more, insulin–resistant DM, ketoacidosis, intercurrent infections and pineal hyperplasia were also present (Rabson and Mendenhall, 1956). The life–span of patients with this syndrome can be longer than one year, during which they develop constant hyperglycaemia followed by diabetic ketoacidosis and death. This condition is accompanied by a constant reduction of insulin levels, which finally becomes insufficient to prevent glucose synthesis by liver and the release of fatty acids by adipocytes (Longo et al., 1999). Until now, more than one hundred disease–causing mutations on *IR* gene have been reported and approximately half of them were identified as type A insulin resistance (Ardon et al., 2014). It has been also reported that mutations with strong effect on the insulin binding are associated with more severe phenotypes, while longer survival rate has been reported for mutations where IR function is partially preserved. Definitive correlations, however, cannot be drawn due to rarity of these conditions and a scarcity of the available experimental data from functional analyses (Longo et al., 2002). The greatest number of the reported mutations in *IR* gene are categorized as missense (64%), followed by nonsense (13%), while the splice site mutations, deletions and insertions in the coding sequence are less common. Most of the mutations that are located in the first 11 exons result in Leprechaunism, while those in the gene responsible for the β–subunit expression are found frequently in patients with Rabson–Mendenhall syndrome (Ardon et al., 2014).

Some variations in *IR* gene are also thought to have influence on the increased risk for the development of type 2 diabetes. Different populations were tested for Val→Met^{985} mutation, giving no conclusive data since only some of these studies support the role of this mutation as a risk factor for diabetes predisposition (Hart et al., 1999), whereas others do not (Hansen et al., 1997). Barroso and his co–workers identified seven single nucleotide polymorphisms in *IR* gene and found that the one present in the sixth intron on the 46th nucleotide base pair (IVS6+43) is associated with increased risk for developing type 2 DM (Barroso et al., 2003).

IR Antibodies

Autoantibodies specific for the insulin receptor were detected as well (Chon et al., 2011; Flier et al., 1975). This rare phenomenon is also called type B insulin resistance and represents a condition where autoantibodies against cell surface proteins are produced. In general, it is similar to Graves' disease, where antibodies against thyrotropin receptor are produced, and myasthenia gravis where antibodies against acetylcholine receptor are produced. Due to this condition, both hyper– and hypoglycaemia may occur with the later being more frequent (Braund et al., 1987). Whether hyper– or hypoglycaemia will occur depends on the consequence of autoantibody binding to IR. If the binding inhibits signal transducing activity, hyperglycaemia occurs, and if the binding stimulates signal transduction, hypoglycaemia occurs (Jeong et al., 2010; Maiza et al., 2013). Type B insulin resistance is characterised by an extreme insulin resistance, dramatic weight loss, severe hyperandrogenism and widespread skin condition known as *acanthosis nigricans*. Medical treatment is usually based on the general immunosuppression, and the one including rituximab, cyclophosphamide and pulse steroids in the form of dexamethasone and methylprednisone gave satisfactory results. Nonetheless, the lack of randomized, placebo–controlled trial is the main disadvantage for testing this treatment approach due to rarity of this disease (Malek et al., 2010).

In one clinical study, which included six non–obese women, severe insulin resistance and *acanthosis nigricans* were detected (Flier et al., 1975). Both basal and glucose stimulated insulin concentrations in plasma of these women were 10– to 100–fold increased. Their response to the injected insulin was markedly reduced, with some of them having to receive 1000 times higher than usual doses of insulin in order to control blood glucose levels. Autoantibodies to insulin were not detected in the blood of these patients and it was proposed that antibody to insulin receptor is responsible for this condition (Flier et al., 1975). In another single case study, patient's auto–anti–IR antibody was associated with hypoinsulinemia. Patient's serum and purified immunoglobulins activated IR β–subunit and IRS–1, which led subsequently to phosphorylation of AKT kinase, thus mimicking insulin hormone. After corticosteroid therapy, both IR autoantibody and the occurrence of hypoglycaemia disappeared (Maiza et al., 2013). In young and adolescent study group, where insulin resistance was already detected, 9.8% of participants had anti–IR antibodies. They had clinical signs of obesity and *acanthosis nigricans*, and no clinical features of autoimmunity. It was suggested that the insulin receptor autoimmunity was responsible for the insulin resistance. These findings highlighted the importance of the examination of the existence of anti–IR antibodies in patients with the childhood onset of insulin resistance, as they represent a considerable number of individuals, and are suitable for the development of new treatment approaches (Zhou et al., 2008).

Two types of antibodies against IR were reported to exist in the blood of persons diagnosed with type 1 and type 2 DM: those that inhibit insulin binding and those that immunoprecipitate solubilized receptors without affecting insulin binding, thus interacting with the regions outside of the insulin binding domain (Batarseh et al., 1988). Predominant types of IR antibodies in patients with DM are of the IgM class, different to anti–insulin antibodies found in patients with insulin resistance, where IgG is a dominant class. Patients with type 2 DM included in the study, who were on insulin treatment and had anti–insulin antibodies, had also anti–IR antibodies. In contrast, other patients with type 2 DM who were also on insulin therapy but did not have anti–insulin antibodies also had no anti–IR antibodies. There was no correlation between anti–insulin and anti–IR antibodies in patients with type 1 DM. A highly significant correlation between the dose of insulin used and anti–IR antibody activity was detected (Batarseh et al., 1988).

In the plasma of patients with type 2 DM or pre–diabetes state, increased protease activity was found compared to healthy persons. Further increase in the proteolytic activity was measured upon high–carbohydrate meal in all three groups. It was suggested that higher proteolytic activity in patients with type 2 DM can cause more intensive cleavage of IR, thus contributing to an increased insulin resistance (Modestino et al., 2019).

HyR in Diabetes

An increase in the HyR abundance would be expected to reduce insulin sensitivity in target tissues, leading to an increased insulin resistance. One study showed that the proportion of HyR was significantly higher in skeletal muscles from patients with type 2 DM than in healthy subjects (Federici et al., 1996). The proportion of HyR was found to correlate with a decrease in both insulin binding affinity and insulin sensitivity, as measured by an insulin tolerance test. When adipose tissue from patients with type 2 DM was analysed, similar results were reported (Federici et al.,

1997). Four studies tested whether the observed alterations in the HyR abundance are primarily defects associated with type 2 DM or represent anomalies associated with other common states of insulin resistance. One study revealed that the amount of IR and HyR was increased in the placenta of the insulin resistant women with gestational hypertension (Valensise et al., 1996). The second study showed that the abundance of HyR was increased in the skeletal muscle of the obese subjects and was correlated to an increased body mass index and decreased insulin sensitivity (Federici et al., 1998a). Greater abundance of HyR was also discovered in the skeletal muscle of patients with insulinoma (Federici et al., 1998b) and correlated with both increased plasma insulin levels and a reduced insulin–mediated glucose uptake (Federici et al., 1998a). The abundance of HyR in the skeletal muscle was compared between lean, glucose–tolerant, non–obese subjects with a different degree of insulin sensitivity, and non–diabetic, overweight subjects (Spampinato et al., 2000). Only in the subjects with lower insulin sensitivity, the HyR/IR ratio was found to be slightly increased (Spampinato et al., 2000). In general, these studies support the hypothesis that changes in the proportion of HyR may contribute to an impaired insulin action in the insulin resistant subjects due to lower affinity of receptors for insulin. It was suggested that an increased expression of HyR may represent a general defect associated with different states of insulin resistance (Sesti et al., 2001).

Acanthosis Nigricans

Acanthosis nigricans, a skin condition associated with dark spots, may be manifested in persons with insulin resistance (both A and B types). At physiological concentrations, insulin regulates carbohydrate, lipid and protein metabolism by binding specifically to IR. On the other hand, at higher concentrations, insulin can promote growth, through binding to IGF–1R (Andersen et al., 2017). Binding of insulin to IGF–1R may stimulate proliferation of keratinocytes and fibroblasts, leading to *acanthosis nigricans* (Phiske, 2014; Cruz and Hud, 1992). Both IGFBP–1 and IGFBP–2 are decreased in obese subjects with hyperinsulinemia, thus increasing plasma concentrations of free IGF–I, which promotes cell growth and differentiation. There are several observations supporting the idea that insulin–dependent activation of IGF–1R can facilitate *acanthosis nigricans* development: (i) IGF-Rs are present in cultured fibroblasts and keratinocytes, (ii) insulin can cross dermoepidermal junction and can stimulate growth and replication of fibroblasts at high concentrations, and (iii) severity of *acanthosis nigricans* in obesity positively correlates with the fasting insulin concentration. Therefore, it is proposed that insulin may promote development of this condition through direct activation of the IGF–I signalling pathway. Areas such as neck and axillae are most often affected, suggesting that sweating and/or friction may be necessary co–events (Phiske, 2014). Although it seemed that the severity of *acanthosis nigricans* correlated with the level of insulin resistance, this is not always the case. In a case study on a Japanese girl who was diagnosed with type A insulin resistance, *acanthosis nigricans* was not observed (Saito–Hakoda et al., 2018). A new hypothesis was made – that IGF–I resistance in skin cells at a receptor or post–receptor level, or even inhibitory action of the mutant IR on IGF–1R signalling, can contribute to this condition.

The Impact of Diabetes on the Cardiovascular System

Cardiovascular complications are the most common complications of diabetes. There is evidence suggesting that the reduction in IGF–1R levels exerts some benefit on the endothelial function in atherosclerosis, although the results are mostly based on animal models (Cubbon et al., 2016). DM is also considered as a hypercoagulable state, since diabetic patients are under an increased risk of the development of thrombosis (Tripodi et al., 2011), a condition involved in approximately 80% of diabetes–related deaths. The concentrations of many clotting factors, such as fibrinogen, thrombin, tissue factors, coagulation factors VII, VIII, XI, XII, kallikrein and von Willebrand factor, are increased in diabetes (Carr, 2001; Vazzana et al., 2012). Besides changes in the concentration, chemical modification (glyco–oxidation) of fibrinogen occurs contributing to thrombosis in diabetes (Gligorijević et al., 2017; Dunn et al., 2006). In general, DM induces increased reactivity of platelets, due to several factors: (i) higher mobilisation of Ca^{2+}, (ii) glycation of the platelet membrane which may lead to increased expression of some receptors, including P–selectin, glycoprotein Ib–IX and fibrinogen receptor IIb/IIIa (Kim et al., 2013; Pomero et al., 2015) and (iii) a decrease in cAMP concentration (Kakouros et al., 2011). It was recently found that platelets from DM2 patients express high levels of the activated P2Y12 receptor (Hu et al., 2017). All these changes lead to an increased sensitivity of platelets to agonists such as thrombin, ADP and collagen, and also to aspirin resistance (Di Minno et al., 2012).

Results on the effect of insulin on the function of platelets are controversial, since some researchers detected insulin–related inhibitory effects, while others found no change at all (Randriamboavonjy and Fleming, 2009; Ferreiro et al., 2010). IGF–I has stimulatory effect on wound healing, alone or in coordination with other molecules. Its impaired interaction with the IGF–1R is believed to cause delayed skin wound repair in patients with DM (Aghdam et al., 2012). Platelets express IR, IGF–1R and HyR at their surface (Hunter and Hers, 2009). Alone, IGF–I is not able to activate platelets but it potentiates their higher activation in the presence of different activators (Kim et al., 2007; Hers, 2007).

IGF–I is present in α granules which are released upon platelet activation, thus increasing local concentration of IGF–I at the site of injury. It was reported that platelets from patients with type 2 DM express higher amount of IGF–1R than platelets from healthy controls, whereas the amount of IR was not significantly affected by DM (Gligorijević et al., 2019). Although it was suspected that insulin has inhibitory effect on platelets (Ferreiro et al., 2010), recent results suggest that there is no correlation (Moore et al., 2015). An increased binding of exogenous IGF–I to platelets potentiates their thrombin–induced aggregation, the effect being more pronounced in the case of platelets derived from patients with DM2 (Gligorijević et al., 2019). According to the obtained results, it seems that the concentration of IGF–I in the circulation may be one of the factors influencing platelet activation in patients with DM. A weak positive correlation between the concentration of HbA1c and the speed of thrombin–induced platelet activation due to exogenous IGF–I implicates that the severity of DM2 may be related to the effect of IGF–I. An increase in IGF–1R may be one of the mechanisms responsible for the observed effect (Gligorijević et al., 2019). In patients with diabetic neuropathy, significantly lower amounts of IGF–1R on erythrocytes were detected compared to control subjects without diabetes and those with diabetes but without neuropathy (Migdalis et al., 1995). Taken all into consideration regarding IGF–1R, it seems that the expression and the role of this receptor in diabetes differ in tissue specific manner. Depending on the tissue, both an increase and a decrease of IGF–I signalling can be detected and the potential effects can be either beneficial or detrimental.

Due to a worldwide increased incidence of diabetes, especially in the developed countries, the attention is drawn to the development of new and effective drugs to either disable transition of glucose intolerance to type 2 DM or to slow down or revert this process. One of the main obstacles in performing this task is the homology between insulin and IGFs, IR and IGF–1R and the overlapping of signalling cascades. It is desirable to construct a molecule being able to bind only to IR, and even more, to interfere only with the metabolic pathway and not the mitogenic. To fulfil this purpose, different orthosteric and antagonistic antybodies have been developed, as well as small peptides, and finaly aptamers (Escribano et al., 2017). Still, a successful solution is far from being found and the thorny road is ahead of scientists trying to untangle this riddle.

Gestational Diabetes

Maternal insulin resistance appears during gestation to ensure enough energy for the foetus. However, when maternal body is not able to encompass these changes, gestational DM (GDM) develops, and its prevalence in Europe is around 5% (Eades et al., 2017). GDM is the type of diabetes diagnosed in the second trimester of pregnancy, and is associated with an increased oxidative stress and inflammation in placenta and foetus (Mrizaki et al., 2014; Radaelli et al., 2003). GDM is associated with different perinatal complications (e.g., macrosomia) which may later in adult life increase a risk of obesity, type 2 diabetes, cardiovascular diseases and metabolic syndrome (Buchanan and Xiang, 2005). This condition is classified as pre–diabetic state as GDM represents an increased risk of developing type 2 DM later in the life (Lee et al., 2008). GDM affects placental structure, altering the transport of nutrients to foetus (Araujo et al., 2015; Brett et al., 2014). Although glucose transport in placenta takes place independently of IR – mainly via GLUT1, the placenta is rich in IR and maternal insulin activates IR signalling pathways, affecting placental metabolism (Hiden et al., 2006). Altered expression of IRs is found in the fetoplacental endothelium of GDM women (Lassance et al., 2013), but IR expression is restored by insulin treatment, establishing normal endothelial function and healthy newborns (Guzmán–Gutiérrez et al., 2014).

In the placenta of obese women, the expression of IGF signalling components and nutrient transporters is altered, and is dependent on the maternal body fat mass, gestational weight gain, and the observed macrosomia (Jansson et al., 2013; Brett et al., 2016; Martino et al., 2016). Increased levels of IGF–I and IGF–II in GDM lead to up–regulation of placental amino acid transporters, resulting in an increase in the placental amino acid levels (Cetin et al., 2005). In obese and mothers with GDM or type 1 diabetes, esterification of fatty acids forms triglycerides which are stored in the placenta (in syncytiotrophoblast) in the form of lipid droplets (Diamant et al., 1982; Elchalal et al., 2005). The key mediators of these actions are fatty acids (Pathmaperuma et al., 2010) and insulin (Hirschmugl et al., 2017). Statins, 3–hydroxy–3–methylglutaryl–coenzyme A reductase inhibitors, are used to lower high levels of the circulating cholesterol, usually diagnosed in obese and women with metabolic syndrome or type 2 DM. Forbes and co–workers (2015) demonstrated that in the first trimester villous tissue explants statins attenuated proliferation induced by IGF–I or IGF–II. Altered IGF–1R distribution in trophoblasts was reported to result from the reduced levels of complex N–glycans, responsible for the expression of the immature receptor at the cell surface (Forbes et al., 2015).

Although insulin secretion increases in pregnancy due to hyperplasia of pancreatic β–cells (Butler et al., 2010), pregnancy is characterised by a relative insulin resistance state, which favours the metabolic needs of the developing foetus, especially in the third trimester (Tan and Tan, 2013). Investigating the effect of pregestational insulin–dependent DM on the content of placental IGF–Rs and IR N–glycans, we found no changes compared to healthy controls (Robajac et al., 2012). The investigated cohort was very small, consisting of only 5 samples in each group, and the pregnant women had regular check–ups and glycaemic control. Our findings implied that a careful control of glucose levels and an adequate therapy enable the maintenance of physiological homeostasis during pregnancy, thus providing normal foetal growth. When considering the number of cases which can be included in the study, one must bear in mind that most pregnancies with uncontrolled diabetes end with either stillbirth or preterm delivery when babies are too small, underdeveloped and with very little chance of survival.

Activation of IR triggers mTOR, a positive regulator of amino acid transport that stimulates cell proliferation and growth. Therefore, a placentomegaly (an increase in the placental size) and foetal macrosomia (overgrowth of the foetus) that are often found in obese and pregnancies complicated with GDM can be a result of mTORC1 signalling (Jansson et al., 2013). The activation of mTOR cascade is related to cell survival, metabolism, growth, proliferation and autophagy and is a sensor of nutrients from the placenta (Jansson et al., 2012). The expression of amino acid carriers is regulated by mTOR pathway. It is important to mention that this pathway can be activated not only by insulin but also by other growth factors. The expression of amino acid transporters can be regulated via mTOR pathway by the levels of amino acids (Roos et al, 2009). This is also found in GDM, where an increased food intake may activate mTOR pathway, further activating amino acid transporters, which can be related to overgrowth and macrosomia (Jansson et al., 2013). Amino acid transporter systems A and L are positively regulated by mTORC1, a critical for the foetal growth (Dimasuay et al., 2016).

As previously stated, some proteins in placenta can interfere with the signalling pathway of insulin, such as Annexin 2 and 14–3–3 proteins, which are overexpressed in GDM placentas and can block the insulin signalling pathway, contributing to insulin resistance found in GDM (Liu et al., 2012). A decrease in the AKT levels (Ruiz–Palacios et al., 2017b) and an increase in the levels of the PI3K subunit p85α (Colomiere et al., 2009; Alonso et al., 2006) occur in the placental insulin resistance in GDM treated with diet. As the insulin signalling pathway is not much damaged in the GDM placenta, it is probable that the insulin resistance can be treated by exogenous insulin therapy (Ruiz–Palacios et al., 2017).

CANCER

Next to cardiovascular, the malignant diseases are the leading cause of mortality in the modern world. Lung, prostate, colorectal, stomach and liver cancer are the most common types of cancer in men, while breast, colorectal, lung, cervix and thyroid cancer are the most common among women. During the last year, 9.6 million people died from conditions associated with cancer (WHO).

Cancer arises as a consequence of neoplastic transformation of cells and tissues – a disturbance of tissue homeostasis in the direction of uncontrolled cell proliferation and inhibition of cell death by apoptosis. Neoplastic transformation is a multistep process, as presented in Figure 4.

Each step is characterised by gene alterations which lead to normal cell transformation into the malignant one. It is known that during the transformation, the cell loses control mechanisms, which leads to an increased growth potential, alterations on cell membranes, karyotype anomalies and various morphological and biochemical changes (Harley et al., 1994).

The traditional view, in which cancer is regarded as a simple lump of cells that uncontrollably divide is discarded, since new findings show that they are complex tissues comprised of many heterogeneous cells which interact in multiple ways. The surrounding of the cancer cells is also very important in tumour genesis and progression, as it plays an active role in the incitement of cancer cell dynamics (Hanahan and Weinberg, 2011).

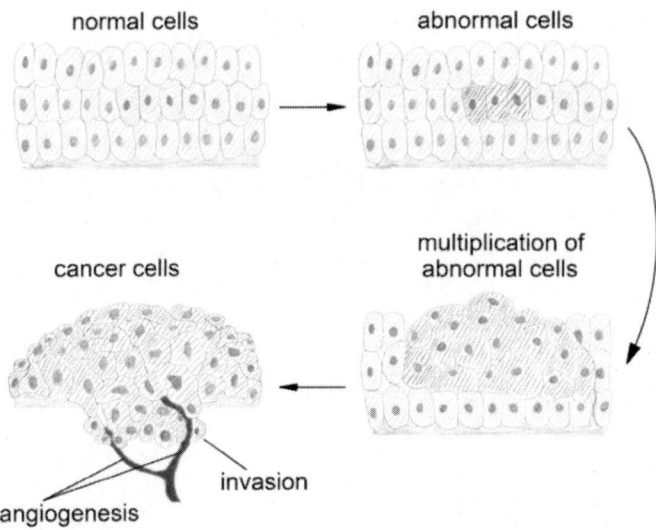

Figure 4. Neoplastic transformation of cells.

General Characteristics of Cancer Cells

Each cancer cell is distinguished by six determinants which enable the cell to evade or circumvent the limiting mechanisms that check that everything is under control. There are ten so–called "hallmarks of cancer" (Fiaschi and Chiarugi, 2012) and each of them will be carefully addressed in the following paragraphs, as cancer has become a hot topic for the last few decades and understanding of its mechanisms is set among the top scientific priorities.

1) *Self–sufficiency of proliferative signals.* One of the best–known characteristics of the cancer cell is its potential for chronic proliferation. Normal tissues possess control mechanisms which carefully "weigh" the amount of growth–promoting and growth–inhibiting signals that are produced, directing cells to some point of the growth/division cycle. Cancer cell acquires transformed phenotype, which is characterised by disordered signalling

cascades, allowing cells to grow and divide uncontrollably (Sever and Brugge, 2015).

Growth signals, which direct normal cell cycle, are usually produced in a paracrine fashion, spatially and temporally conditioned, and often "stored" in pericellular space, where their action is controlled by a multitude of enzymes which activate them in a tissue localized manner. Experimental study of this kind of regulation is demanding and hardly accessible, which is the reason why we do not know much about that level of regulatory mechanisms (Kataoka et al., 2003).

On the other hand, cancer cells pave their own way through the "wilderness," without letting neighbouring cells decide on their faith. Growth factors which they produce act in an autocrine manner, binding to surface receptors which are either structurally altered to become more sensitive to ligand binding and/or are produced in greater quantities which makes them more responsive to limiting amounts of ligands (Walsh et al., 1991). Beside altered upstream members of the signalling cascade (ligands and receptors), the downstream elements can also be changed becoming hyperactive, usually through somatic mutations that render them constitutively active. One of the downstream signalling molecules that is very often mutated in cancer cells is the catalytic subunit of PI3K (Jiang and Liu, 2009; Yuan and Cantley, 2008).

2) *Insensitivity to anti–proliferative signals.* One of the ways to control cell growth and division is through negative–feedback mechanisms. Consequently, the cell can respond to various signals to which it is exposed (Amit et al., 2007). Any disturbance in the circuit can lead to continuous proliferation. It is documented that in various types of cancer, some stage of this control becomes defective. For instance, mTOR is one of the key regulators that influence the flux of information in the cell, upstream and downstream of the PI3K pathway. When this switch gets hampered, the loss of the negative–feedback mechanism leads to a

prolonged activation of PI3K and its effector AKT/PKB, thus disabling the anti–proliferative role of mTOR (Sudarsanam et al., 2010).

Although an excess signalling leads to cellular proliferation, it was shown that overexpression of oncogenes, leads to a state called senescence and/or apoptosis. Mooi and Peeper (2006) found cells with morphological features of senescence in some cases of melanoma, so it seems that this is an obstacle that transformed cell must overcome in order to achieve pro–proliferative characteristics.

Tumour–suppressor genes are another level of protection from unconstrained cell cycle. They have an ability to abort the progression of a cell through cyclic changes if they "sense" something is wrong, by integration of signals from extracellular and intracellular sources. One of the prototypical tumour–suppressor genes is the one that expresses Rb (retinoblastoma–associated) (Burkhart and Sage, 2008) and TP53 proteins (Soussi and Wiman, 2015). Rb protein usually filters the signals from the extracellular milieu, while TP53 monitors the intracellular changes, such as the supply of oxygen and nutrients, changes in genome structure etc. When the messages which cell receives are unfavourable, the programs for senescence or apoptosis are activated. If the expression of these proteins is somehow inhibited, the cell is left without break and is free to grow and divide without any limitation.

Another way in which proliferative potential of the cell is kept under control is through contact inhibition, which prevents cells to migrate and divide, when in contact with the neighbouring cells. Cancer cells are insensitive to this mechanism and will keep moving across adjacent cells and dividing, forming disordered, multi–layered patterns. Since contact inhibition is preserved in normal cells, they will stop dividing in the contact with the cancer cells. That can be one of the mechanisms by which cancer affects

distant tissue it colonises during metastasis, where it leads to shrinkage of the affected organ (Ribatti, 2017).
3) *Invasion and metastasis*. After some time, proliferating cancer cells start to populate surrounding tissues, as a single cell or as a group of cells interconnected by adhesion molecules and other communication junctions. The first step in this process is detachment of the cell from its environment e.g., other cells and extracellular matrix, followed by the change in its shape, enabling it to overcome spatial restrictions and to migrate without constraint. Cancer cells develop a phenotype that is a feature of embryonic tissues when cells actively migrate and proliferate in the process of organogenesis (Berx and van Roy, 2009).

Invasion is the first step in the cascade of events, with the final act being dissemination to distant, receptive tissues (metastasis). The second step in the metastasis process is the intravasation, when cells escape the primary site of origin, and enter the circulatory system through local blood and lymphatic vessels. When they reach certain organ, the cell(s) leave the vessels lumina (extravasion), and populate it, forming small clusters which grow into a micrometastatic lesions and finally to macroscopic tumour (colonization) (Seyfried and Huysentruyt, 2013). What is interesting is that certain tumour cells prefer to colonize only specified organs (the "seed and soil" theory). It is believed that cancer cells can thrive only in the environment that is similar to the place where primary tumour developed, so, for example, prostate cancer usually metastasizes to bones; colon cancer has a tendency to metastasize to the liver, while in women stomach cancer often metastasises to the ovary (Akhtar et al., 2019).

The cancer cells' environment is not just a silent observer in the process of metastasis, but an active helper. Cancer cells manipulate the normal cells around them, by sending signals which activate them. It is known that macrophages at the periphery of tumour, after activation by cancer cells, secrete proteases which degrade the extracellular matrix, helping cancer cells in invasion and colonization (Kessenbrock et al., 2010). This type of metastasis is termed "mesenchymal," and is characteristic of cancers that arise from epithelial tissues (carcinomas). There are two other ways in which cancer cells can invade and colonize distant tissues. One way is called "collective" invasion and involves a set of cells which travel together into adjacent tissues, and the other is "amoeboid" invasion, in which cell takes more plastic form, that enables it to squeeze through cracks in the extracellular matrix, rather than degrade it (Krakhmal et al., 2015).

4) *Limitless replication*. The main characteristic of a normal cell is its mortality. After a certain number of growth/division cycles the cell enters a state in which it is viable, but without the possibility to replicate, the so called senescence. If the cell goes on beyond that point, it enters a crisis and succumbs to cellular death, apoptosis. These two points (senescence and crisis) represent checkpoints which prevent cells to enter a limitless replication sequence. One of the structures which prevent this are small repeats at the end of chromosomes, called telomeres. After each division cycle, a telomere shortens slightly, eventually losing its ability to protect the chromosome from end–to–end fusion, resulting in a disruption of karyotype stability, thus triggering a program of conducted cellular death. Cancer cells, however, gain possibility to overcome this hurdle. Their telomeres do not shorten, enabling them to become immortal. The majority of cancer cells produce a telomerase, which extends telomeric DNA after division, making cells prone to endless division (Jafri et al., 2016).

5) *Continuous angiogenesis*. Like any other cell, the cancer cell also needs nutrients, oxygen and waste disposal system to survive. These needs are satisfied via blood vessels, which deliver nutrients and oxygen and remove carbon–dioxide and metabolic waste products. During embryogenesis, blood vessels are continuously formed. Endothelial cells are differentiated into capillaries and bigger blood vessels, and new sprouts of capillaries are formed from the existing, making the circulatory system of the body

(Folkman, 2007). During adulthood, the angiogenesis occurs rarely, only in specific cases such as wound healing and during female reproductive cycle (Tepper et al., 2005). The occurrence of intensive angiogenesis is noticed in neoplastic transformation of cells. As cancer cells divide and grow they get further and further away from the nearest capillary and the source of vital components dries out. The locally induced hypoxia can activate the expression of vascular endothelial growth factor (VEGF), pro–angiogenic factor that stimulates the formation of new blood vessels (Katayama et al., 2019). Oncogenes can also activate the expression of this protein. A number of angiogenic inhibitors can be detected in the circulation. They are usually fragments of proteins that are not inhibitors themselves (plasmin, collagen). The inactivation of these entities can lead to neoplastic transformation in cells (Ribatti, 2009a).

6) *Escaping apoptosis*. Programmed cell death, or apoptosis is a mechanism through which cell commits voluntary suicide, and it happens when proper functioning is irreparable and if there are no other possibilities for rescue. The most manifested stressors which induce this mechanism are disturbances in signalling (survival signals) and the anomalies in the DNA. The cellular fate is determined by fine balance between pro– and anti–apoptotic factors which belong to the Bcl–2 family. The main pro–apoptotic proteins are Bax and Bak, and they are found in the outer mitochondrial membrane. Three main anti–apoptotic proteins are Bcl–2, Bcl–x_L and Bcl–w (Adams and Cory, 2007).

As previously said, TP53 is one of the main tumour supressors in the cell, and its activation induces a signalling cascade which leads to apoptosis. Cancer cells evolve a mechanism to inactivate this protein, thus, eliminating a critical sensor (Zawacka–Pankau and Selivanova, 2015). They can also produce a pro–surviving factors, such as IGFs (Yu and Rohan, 2000), which can deceive the cell disabling its apoptosis. Down–regulation of pro–apoptotic proteins is another strategy for survival (Igney and Krammer, 2002).

7) *The change in cell metabolism.* When supplied with enough oxygen, cells synthesize ATP through a process of oxidative phosphorylation: glycolysis is executed in the cytosol, glucose is transformed to pyruvate and then to carbon dioxide in mitochondria. Under anaerobic conditions, the pyruvate is not directed to mitochondria and the amount of ATP produced per 1 mol of glucose is very low. German physiologist and doctor Otto Warburg noticed that cancer cells prefer to produce ATP solely under the glycolytic condition, the state being called "aerobic glycolysis" (Warburg, 1956). In cancerous tissue, as a result of inflammation, availability of nutrients and hence the glucose are reduced. The metabolic shift is partially achieved through an up–regulation of GLUT1 transporters which increase the influx of glucose in the cell. Activated oncogenes are among the main stimulators of glycolysis (Hsu and Sabatini, 2008). It is not known why cancer cells use aerobic glycolysis mechanism for creating energy, especially since this production is nutritionally costly. One of plausible explanations is that, in this way, the intermediates of the path are redirected towards synthesis of important molecules such as nucleotides and certain amino acids (Vander Heiden et al., 2009).

8) *Escaping immune destruction.* The immune system monitors every irregularity and destroys cells which are not inherent to an organism, or which are in some way transformed. This is a way to halt any process which leads to destructive changes. The cells of innate and adaptive immune system constantly patrol through the body, and if they detect changed structures, they destroy the affected cell. Tumour and, eventually cancer, arises as a consequence of transformed cells managing to escape the immune system surveillance. Cancer cells possess the ability to modify their antigens in order to trick the immune system, and survive. The cells of the immune system can destroy certain portion of transformed cells, but a small number of them, which acquired favourable mutations, remain and become dominant over the

immune system, by establishing immunosuppressive environment (Ribatti, 2017b).

9) *Genome instability and mutation.* Cancer cells are prone to high mutation rate, where subpopulation of cells achieves a favourable phenotype, which enables them to gain a selective advantage and thrive (Negrini et al., 2010). This genomic instability arises as a consequence of increased sensitivity to mutagenic signals and inactivation of proteins which take part in the maintenance of the genome. These proteins detect and repair mistakes made during genomic expression, and neutralise potentially damaging agents (Jackson and Bartek, 2009). Increased rate of general mutation is also a consequence of mutational inactivation of tumour suppressor genes, which monitor the activities inside the cell and, if necessary, direct cell to senescence or apoptosis, such as *TP53* (Lane, 1992).

10) *Tumour–promoting inflammation.* The constituent parts of tumour tissue are immune cells, equally originating from innate and adaptive arm of the immune system (Man et al., 2013). The most obvious explanation for their appearance is the attempt of the immune system to eradicate tumour cells. By trying to eliminate the transformed cells, they secrete a variety of signalling molecules, which is similar to the immune response to invading pathogens or in wound healing. But these molecules, which should help healthy cells, this time, do more harm than good. It was previously said that cancer cells gain the advantage by escaping the immune surveillance. When they do so, they are free to replicate and proliferate, helped by the stimulatory molecules that immune cells secrete (various growth factors, such as EGF and VEGF, pro–angiogenic factors, chemokines) (Grivennikov et al., 2010).

IGF System and Cancer

The IGF system is a very important mediator which orchestrates cellular actions that lead to growth, division, differentiation or programmed cellular death. It must be tightly controlled, since any mistake can potentially lead to disturbances which, in the case of cancer, may end up with uncontrolled cell division (Samani et al., 2007).

An increased expression of IGF–I, IGF–II, IGF–1R or the combination of these molecules was found in many types of tumours such as glioblastoma (Sandberg et al., 1988), neuroblastoma (Lichtor er al., 1993), breast cancer (Gebauer et al., 1998), colorectal (Freier et al., 1999), pancreatic (Bergmann et al., 1995) and ovarian cancer (Sayer et al., 2005). The elements of the IGF system act as paracrine and/or autocrine factors in the stimulation of tumour growth *in situ* or during tumour progression, and their roles depend upon type of the tissue. As mentioned earlier, the *IGF–II* gene is expressed only from the paternal allele, through a mechanism known as imprinting, which silences the mother's copy of the gene. The loss of imprinting activates both alleles, and raises significantly the amount of IGF–II in the circulation. An increased expression of IGF–II was noted in Wilms' tumour (Ogawa et al., 1993) and subsequently in various gynaecological (Chen et al., 2000) and gastrointestinal neoplasia (Cui et al., 2003).

The action of IGFs is restrained, as they are bound to IGFBPs which control their action, i.e., restrict them from binding to IGF–1R. The change in the amount of these proteins can influence the amount of IGF bound to a cognate receptor (Clemmons et al., 1995). IGFBP–3 is the most abundant binding protein, so the variation in its concentration exerts the greatest influence on the availability of IGFs. *In vitro* experiments have shown that IGFBP–3 plays a protective role against cancer, through direct binding of IGFs or through IGF independent actions, since it can directly interact with specific receptors on the surface of the cell or inside it, and activate the apoptotic cascade (Grimberg, 2000). Anti–apoptotic properties of IGFBP–3 were so far confirmed only in *in vitro* studies (Grill et al., 2002), the results being inconsistent in epidemiological studies (Renehan et al., 2004).

IGFBPs can also have pro–tumorigenic actions, by expressing their IGF independent effects. It was shown that IGFBP–2 can inhibit the action of PTEN, a major tumour suppressor (Zeng et al., 2015), and the positive correlation between serum IGFBP–2 concentration and cancer cell invasiveness was found in patients with colorectal (Šunderić et al., 2014), prostate (Uzoh et al., 2011), breast cancer (So et al., 2008) and various types of leukaemia (Kühnl et al., 2011).

Metastasis is a process which includes interplay between the cancer cell and its surrounding. In order to leave its original habitat and invade other regions, the cell must surpass many barriers, and usually not so welcome microenvironment. Some of the obstacles it must overcome are (Chambers et al., 2002): (i) the need for nutrients and oxygen, (ii) the existence of extracellular matrix, (iii) tissue barriers and (iv) the adaptation for survival in a new environment. The elements of the IGF system can provide the cell a much needed help in any of these steps.

The main characteristic of the cancer cell is its fast rate of growth and proliferation. In order to keep up with such hectic pace, the cell's metabolism craves for nutrients and oxygen. As the cell grows and divides, it becomes more and more remote from the nearest capillary as a source of nutrient/fuel, so it must overcome this problem. One of the solutions is to secrete factors which stimulate the formation of new blood vessels in the process known as neovascularisation. The main trigger for this stimulation can be local hypoxia. IGFs and insulin can precede and/or augment the hypoxic stimulus (Samani et al., 2007). The experiments with cultured cells demonstrated that IGF–I/II can induce the expression of hypoxia–inducible factor 1α (HIF–1α) (Feldser et al., 1999), which can activate the HIF–1/arylhydrocarbon receptor nuclear complex, involved in the activation of genes that contain the hypoxia response element, such as VEGF, a major tumour–induced pro–angiogenic factor (Zelzer et al., 1998). IGFs act via the MAPK and PI3K pathways. The other way in which IGF peptides can stimulate angiogenesis is through direct stimulation of endothelial cells to migrate and differentiate. The IGFs play a paracrine role by entering the surrounding capillaries, where they participate in cell survival and stability (Grulich–Henn et al., 2002). The hypoxia can, on the other hand, induce the production of IGFs. The experiments performed on the human hepatocellular carcinoma cell line (HepG2), confirmed that IGF–II production is up–regulated by hypoxia, through involvment of Erg–1 transcription factor and Wilms' tumour (WT) 1 suppressor gene. Hypoxia activates Erg–1 and induces the expression of IGF–II promoter gene, while the WT1 suppressor gene is silenced (Kim et al., 1998).

If they want to leave the place of origin, cancer cells must break through a network of intertwined threads which constitute the extracellular matrix. In order to do so, they must degrade the matrix by the action of several enzymes called matrix metalloproteinases (MMPs), especially MMP–2 and MMP–9 which are associated with angiogenesis and tumour invasion. It was shown that the elements of the IGF system can regulate the expression of these MMPs, thus promoting cell migration and invasion (Long et al., 1998; Mira et al., 1999).

IGF–1R/IR and Cancer

Biological actions of IGF peptides, as already said, are achieved through binding to three types of receptors: IR, IGF–1R and HyR (IR/IGF–1R). In adult, healthy tissues, IR is responsible for signals that stimulate metabolic actions, while IGF–1R transfers the information that stimulates cellular growth and division. In transformed cells, this is not so clear–cut, as the signalling pathways can become interconected and synergistic (Malaguarnera and Belfiore, 2011). Cancer cells usally overexpress IR–A isoform by 60% to even 100% compared to healthy cells (Malaguarnera and Belfiore, 2011). Since this type of receptor has high binding affinity for IGF–II and IGF–1R for IGF–I and IGF–II, stimulatory effect of IGFs can be maximized.

IGF–2R acts as a tumour suppressor since it binds IGF–II leading it to degradation inside the cell. Various structural aberrations, such as mutations, loss of heterozygosity and microsatellite instability, can alter the activity of this receptor, as was found in many tumours (Souza et al., 1999). Although the primary function of IGF–2R is to bind and sequester IGF–II from the extracellular space, it was discovered that it can activate

TGF–β1 (O'Gorman et al., 2002), a growth inhibitor for most types of cells and influence the sensitivity of cancer cells to host immune system (Motyka et al., 2000).

Activation of IGF–1R leads to a series of cascade events, with a main goal to save the cell from apoptosis and to stimulate its growth and division (Figure 5). Binding of IGF–I/II to IGF–1R activates PI3K–AKT–mTOR pathway, which results in protein synthesis, growth and preparation of the cell to replicate (Butler et al., 1998).

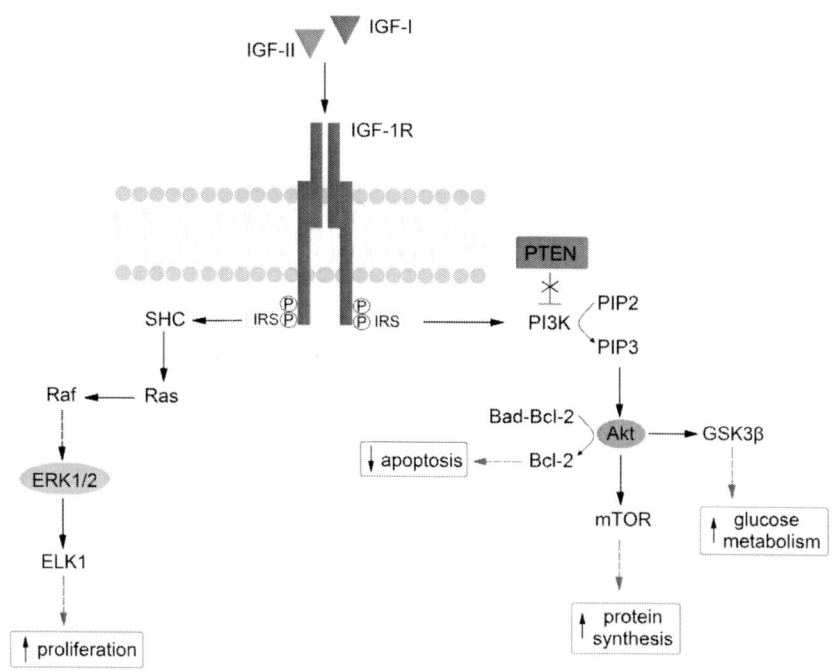

Figure 5. IGF–1R signaling pathway in cancer.

The activity of PI3K is controlled by a PTEN, a well–known tumour suppressor. It acts as phosphatase, causing dephosphorylation of PI3K, thus inhibiting downstream signalling process. The PTEN molecule is the second most commonly deactivated tumour–suppressor in human cancer. Activation of PI3K–AKT pathway enables the cell to avoid pro–apoptotic signals by inactivation of the pro–apoptotic Bcl–2–associated agonist of cell death, known as Bad protein (Moorehead et al., 2003).

Reactive oxygen species (ROS) are by–products of the cell metabolism, arising from processes responsible for the energy production, cell signalling and destruction of foreign "invaders," among other processes. The cell holds an intrinsic balance which is based on a delicate equilibrium between produced ROS and molecules that act as their neutralizers (antioxidants). When this state becomes disrupted, the cell enters a condition which is called an oxidative stress (Betteridge, 2000). Oxidative stress usually has devastating effect on a cell, by inducing its death or senescence at least, but that's not always the case. Generated ROS can induce positive reinforcement on each step the cell needs in order to transform itself to a malignant form (Reuter et al., 2010). One of the first steps in this transformation cycle is the so called "epithelial–mesenchymal" (EMT) transition, in which a cell gains properties which enable it to acquire a more aggressive phenotype. ROS can influence a number of signalling pathways that are involved in this process, particularly those which involve an extracellular matrix (ECM) (Mani et al., 2008).

It is already known that a change in the glycosylation pattern accompanies the shift towards neoplastic transformation. Cancer cells often exhibit novel glycan structures, which are not present in normal cells; however, many of them are being expressed in the foetal phase (Dennis and Laferte, 1987). Each step of the malignant transformation is followed by some alteration in the glycan structure. Inside the transformed cell, the expression levels of certain glycosyl–transferases are altered, in order to encourage metastasis and cancer spreading (Gu et al., 2009). It is known that some tumour types are characterised by altered glycosylation patterns, and this feature can serve to establish biomarkers, if further clinical investigations show their tissue specificity (Adamczyk et al., 2012). IGF–1R is glycosylated and possesses a unique glycan signature, which enables proper functioning, reflected in ligand binding and signal transduction. The change in the glycosylation pattern can influence these characteristics and alter the responsiveness of the receptor to its ligands.

IGF System and Colorectal Cancer

Colorectal cancer is the third most common cancer diagnosed in the general population. In 2018, there were over 1.8 million new cases worldwide. Out of 25 countries with the highest incidence rate of this type of cancer, Hungary leads, with 51.2 new cases per 100.000 people, in both sexes (World Cancer Research Fund/American Institute for Cancer Research).

Clorectal cancer is derived from the epithelial cells that coat the inner surface of the colon and rectum (De Rosa et al., 2015). The initial investigations, which consisted of immunostaining of the higher–grade and higher–stage cancer tissues, showed strong response to anti–IGF–1R antibody, in adenocarcinomas and their metastases (Hakam et al., 1999). The colorectal cancer cell lines (CaCo–2) demonstrated higher expression rate of IGF–II and IGF–1R (Zarrilli et al., 1994).

Although the preclinical data is convincing, the epidemiological studies are not so straightforward. The study on UK patients with acromegaly revealed no higher risk for the occurrence of colorectal cancer, than in the healthy population, although they have increased level of IGF–I in the circulation (Orme et al., 1998). On the other hand, in a prospective study conducted on 210 colorectal cancer patients that underwent resection, immunohistochemical analysis showed that the expression of IGF–I and IGF–1R seems to increase with tumour size (Shiratsuchi et al., 2011).

Post–Translational Changes of IGF Receptors in Colorectal Cancer (An Experimental Model)

Post–translational modifications (PTM) represent a tool by which the structure and the function of proteins can be manipulated, exerting a fine tuning control of proteins and enabling them to change their activity in accordance with the changing environmental conditions. The majority of PTMs are enzymatically controlled and coordinated, but sometimes proteins can be modified randomly, in an unplanned manner (Walsh et al., 2005). As a consequence, they can even lose their function, or gain a new one, such as antigenicity (Kurien et al., 2006).

One such modification is protein oxidation, caused by ROS. Cancer is a state with elevated oxidative stress, which reflects on proteins as one of the targets. By employing the radioreceptor assay, we showed that ^{125}I–IGF–I was able to bind more strongly to IGF–1R in tumour samples than in non–tumour, whereas the binding of ^{125}I–IGF–II to IGF–1/2R was the same in both groups of samples (Nedić et al., 2013). Although oxidative stress accompanies malignant processes in the cell and its environment, not all proteins are equally exposed or susceptible to modification. Whether proteins will be oxidatively modified depends on their structure, location, concentration and the nature, as well as on the concentration and type of ROS present (Barelli et al., 2008). DNP modification of membrane proteins isolated from colon tissue revealed that total membrane proteins are only slightly modified in tumour samples (Nedić et al., 2013). Unlike oxidized IGF–2R, the oxidized IGF–1R could be detected only in fragments, suggesting that oxidation of IGF–1R makes it more susceptible to degradation than oxidation of IGF–2R. An increased binding of ^{125}I–IGF–I to IGF–1R in tumour samples, without increased concentration of the receptor compared to healthy tissues, led to a conclusion that structural changes induced an increase in affinity, which may be a result of the oxidative alteration of the protein (Nedić et al., 2013).

Aberrant glycosylation is a common feature of neoplastic changes in the cell. The glycan moiety is mostly found on cell surface and extracellular matrix, and participates in cellular interactions (Ohtsubo and Marth, 2006). Since the cell–environment contact is very important in cancer spreading, changes in the glycan structure of proteins can be more predictive than changes in the structure of the protein itself. As already mentioned, IGF–Rs are glycosylated and IGF–1R is overactive in cancer pathology. Until now, there is no data on the glycosylation pattern of IGF receptors in colon cancer. In an unpublished study (Robajac et al., 2019), the use of the reverse–phase lectin microarray revealed the strongest interaction of IGF–Rs isolated from colon cancer tissue and adjacent non–tumour tissue, with lectins MAL–II and WGA, implying that receptors are rich in terminal α2,3 Sia (MAL–II) and GlcNAc β1,4 GlcNAc or Sia (WGA). Our preliminary results with lectins ConA and AAL suggest that

IGF–Rs in tumour tissues are exposed to an increased fucosylation and mannosylation; both of these glycosylation changes are already known to accompany cell transformation to cancer phenotype. The glycans of IGF–Rs most probably affect their structure, ligand binding, signalling, and interaction with physiological lectins.

IGF in Therapy

Radio–/chemotherapy is still a primary choice for treating cancer. Although cells are exposed to agents which act as cytostatics (unfortunately, healthy cells are also affected), after some time, cancer cells can escape the inhibitory effect of therapy and become resistant, or even appear as more aggressive phenotype (Jones, 2016). IGF–1R plays a significant role in acquiring such features. Some of the mechanisms involved in the resistance to therapy are: (i) loss of the inhibitory effect of a number of molecules such as WT1 and miRNA on IGF–1R expression (Chen et al., 2011; Wang et al., 2014), (ii) overactivation of IGF–1R by constitutively secreted IGF–I (Montazami et al., 2015) and (iii) overactivation of IGF–1R by Src oncogene (Werner and Le Roith, 2000). Signals originating from IGF–1R control the expression of the multidrug resistance protein 1 (MDR1) (Benabbou et al., 2014), a member of ABC (ATP–binding cassette transporter) protein family, which act as importers and exporters of different molecules in a cell. Their elevated expression in cancer leads to an increased efflux of cytostatic drugs and consequential decrease in the therapeutic response (Zhang et al., 2015). In addition to residing in the cell membrane, IGF–1R can be translocated into the nucleus where it acts as a transcriptional regulator, by controlling the expression of genes responsible for the cell cycle progression (Sarfstein et al., 2012; Vesel et al., 2017).

There have been trials to design a therapeutic medium which will interfere with IR/IGF–1R signalling, in order to silence the signalling cascade that stimulates cancer cells to divide in an uncontroled manner. The first targets are IGF peptides. Chen and his co–workers (2018) designed an antibody that neutralised both IGF–I and IGF–II, preventing them to bind to IGF–1R. This antibody inhibited the growth of the breast cancer cell line by as much as 40%, at the lowest concentration added to the culture medium. Another direction of intervention is to use anti–receptor antibodies, which would compete with IGFs for binding to IGF–1R, but without activation of the signalling cascade. Investigations of that type have even reached the phase III of clinical investigation (figitumumab), but the outcome was rather disappointing. Anti–receptor antibodies were not efficient enough and caused metabolic toxicity (Langer et al., 2014). The problem with anti–receptor antibodies was that, although designed not to interact with IR, they nevertheless caused hyperinsulinemia and hyperglycaemia. The explanation can be found in the activation of the feedback mechanism, since the pituitary gland senses insufficient IGF activity and secretes higher amounts of growth hormone. Growth hormone causes depression of insulin sensitivity in tissues, which ultimately leads to hyperinsulinemia, and consequently to hyperglycaemia (Pollak, 2012).

CONCLUSION

Two very distinct but profoundly related processes of placental development and carcinogenesis have been used as model systems throughout this chapter to explain the roles of membrane receptors. As discussed, they are both characterised by fast cell growth and proliferation, high invasiveness and cell migration. One of them mimics tumour metastasis while the other one is an actual process of tumour metastasis, the only difference being tight regulation of oncogenes and oncogenic processes during placentation, as extensively reviewed by West et al. (2018). We examined the role of IGF–1R, IR and IGF–2R in these processes, as these membrane glycoproteins are responsible for glucose metabolism, protein and nucleic acid synthesis, cell growth, proliferation, survival, apoptosis. As explained in this chapter, all of these closely related and to some extent opposite processes intertwine and the plethora of overlapping signals settles the IGF system at the crossroad between the good (desirable) and bad (harmful) pathophysiological events. It is the surrounding and, hence, the phenotype that is the driving force to reach the final verdict on whether the story will be bitter or sweet.

ACKNOWLEDGMENTS

This work was supported by the Ministry of Education, Science and Technological Development of the Republic of Serbia, under the project number 173042. Data reported in this chapter resulted from several national and bilateral projects.

REFERENCES

Adamczyk B, Tharmalingam T, Rudd PM. Glycans as cancer biomarkers. *Biochim Biophys Acta*. 2012;1820:1347–1353.

Adams JM, Cory S. The Bcl–2 apoptotic switch in cancer development and therapy. *Oncogene*. 2007;26:1324–1337.

Aghdam SY, Eming SA, Willenborg S, Neuhaus B, Niessen CM, Partridge L, Krieg T, Bruning JC. Vascular endothelial insulin/IGF–1 signaling controls skin wound vascularization. *Biochem Biophys Res Comm*. 2012;421:197–202.

Akhtar M, Haider A, Rashid S, Al–Nabet ADMH. Paget's "Seed and Soil" theory of cancer metastasis: An idea whose time has come. *Adv Anat Pathol*. 2019;26:69–74.

Alonso A, Del Rey CG, Navarro A, Tolivia J, Gonzalez CG. Effects of gestational diabetes mellitus on proteins implicated in insulin signalling in human placenta. *Gynecol Endocrinol*. 2006;22:526–535.

Amit I, Citri A, Shay T, Lu Y, Katz M, Zhang F, Tarcic G, Siwak D, Lahad J, Jacob–Hirsch J, et al. A module of negative feedback regulators defines growth factor signaling. *Nat Genet*. 2007;39:503–512.

Andersen M, Nørgaard–Pedersen D, Brandt J, Pettersson I, Slaaby R. IGF1 and IGF2 specificities to the two insulin receptor isoforms are determined by insulin receptor amino acid 718. *PLoS One*. 2017;12:e0178885.

Anderson CM, Henry RR, Knudson PE, Olefsky JM, Webster JG. Relative expression of insulin receptor isoforms does not differ in lean, obese, and noninsulin–dependent diabetes mellitus subjects. *J Clin Endocrinol Metab*. 1993;76:1380–1382.

Anisimov VN, Bartke A. The key role of growth hormone–insulin–IGF–I signaling in ageing and cancer. *Crit Rev Oncol/Hematol*. 2013;87:201–223.

Annunziata M, Granata R, Ghigo E. The IGF system. *Acta Diabetol*. 2011;48:1–9.

Aplin JD, Lacey H, Haigh T, Jones CJ, Chen CP, Westwood M. Growth factor–extracellular matrix synergy in the control of trophoblast invasion. *Biochem Soc Trans*. 2000;28:199–202.

Arabkhari M, Bunda S, Wang Y, Wang A, Pshezhetsky AV, Hinek A. Desialylation of insulin receptors and IGF–1 receptors by neuraminidase–1 controls the net proliferative response of L6 myoblasts to insulin. *Glycobiology*. 2010;20:603–16.

Araujo JR, Keating E, Martel F. Impact of gestational diabetes mellitus in the maternal–to–fetal transport of nutrients. *Curr Diab Rep*. 2015;15:569.

Ardon O, Procter M, Tvrdik T, Longo N, Mao R. Sequencing analysis of insulin receptor defects and detection of two novel mutations in *INSR* gene. *Mol Genet Metab Rep*. 2014;1:71–84.

Bach LA. IGF–binding proteins. *J Mol Endocrinol*. 2018;61:T11–T28.

Barelli S, Canellini G, Thadikkaran L, Crettaz D, Quadroni M, Rossier JS, Tissot JD, Lion N. Oxidation of proteins: Basic principles and perspectives for blood proteomics. *Proteomics Clin Appl*. 2008;2:142–157.

Barroso I, Luan J, Middelberg RPS, Harding AH, Franks PW, Jakes RW, Clayton D, Schafer AJ, O'Rahilly S, Wareham NJ. Candidate gene association study in type 2 diabetes indicates a role for genes involved in β–cell function as well as insulin action. *PloS Biology*. 2003;1:41–55.

Batarseh H, Thompson RA, Odugbesan O, Barnett AH. Insulin receptor antibodies in diabetes mellitus. *Clin Exp Immunol*. 1988;71:85–90.

Belfiore A, Malaguarnera R, Vella V, Lowrence MC, Sciacca L, Frasca F, Morrione A, Vigneri R. Insulin receptor isoforms in physiology and disease: an updated view. *Endocr Rev*. 2017;38:379–431.

Benabbou N, Mirshahi P, Bordu C, Faussat AM, Tang R, Therwath A, Soria J, Marie JP, Mirshahi M. A subset of bone marrow stromal cells regulate ATP–binding cassette gene expression via insulin–like growth factor–I in a leukemia cell line. *Int J Oncol*. 2014;45:1372–1380.

Benecke H, Flier JS, Moller DE. Alternatively spliced variants of the insulin receptor protein. Expression in normal and diabetic human tissues. *J Clin Invest*. 1992;89:2066–2070.

Benyoucef S, Surinya KH, Hadaschik D, Siddle K. Characterization of insulin/IGF hybrid receptors: contributions of the insulin receptor L2 and Fn1 domains and the alternatively spliced exon 11 sequence to ligand binding and receptor activation. *Biochem J*. 2007;403:603–613.

Bergmann U, Funatomi H, Yokoyama M, Beger HG, Korc M. Insulin–like growth factor I overexpression in human pancreatic cancer: evidence for autocrine and paracrine roles. *Cancer Res*. 1995;55:2007–2011.

Berx G, van Roy F. Involvement of members of the cadherin superfamily in cancer. *Cold Spring Harb Perspect Biol*. 2009;1:a003129.

Betteridge DJ. What is oxidative stress? *Metabolism*. 2000;49(2 Suppl 1):3–8.

Bhaumick B, George D, Bala RM. Potentiation of epidermal growth factor–induced differentiation of cultured human placental cells by insulin–like growth factor–I. *J Clin Endocrinol Metab*. 1992;74:1005–1011.

Blum WF, Alherbish A, Alsagheir A, El Awwa A, Kaplan W, Koledova E, Savage M. The growth hormone–insulin–like growth factor–I axis in the diagnosis and treatment of growth disorders. *Endocr Connect*. 2018;7:R212–R222.

Brahmkhatri VP, Prasanna C, Atreya HS. Insulin–like growth factor system in cancer: novel targeted therapies. *Biomed Res Int.* 2015;538019.

Braund WJ, Williamson DH, Clark A, Naylor BA, Buley ID, Chapel HM. Autoimmunity to insulin receptor and hypoglycaemia in patient with Hodgkin's disease. *Lancet.* 1987;1:237–240.

Brett KE, Ferraro ZM, Holcik M, Adamo KB. Placenta nutrient transport–related gene expression: the impact of maternal obesity and excessive gestational weight gain. *J Matern Fetal Neonatal Med.* 2016;29:1399–1405.

Brett KE, Ferraro ZM, Yockell–Lelievre J, Gruslin A, Adamo KB. Maternal–fetal nutrient transport in pregnancy pathologies: The role of the placenta. *Int J Mol Sci.* 2014;15:16153–16185.

Brioude F, Kalish JM, Mussa A, Foster AC, Bliek J, Ferrero GB, Boonen SE, Cole T, Baker R, Bertoletti M, et al. Expert consensus document: Clinical and molecular diagnosis, screening and management of Beckwith–Wiedemann syndrome: an international consensus statement. *Nat Rev Endocrinol.* 2018;14:229–249.

Brugts MP, van Duijn CM, Hofland LJ, Witteman JC, Lamberts SW, Janssen JA. IGF–I bioactivity in an elderly population: Relation to insulin sensitivity, insulin levels, and the metabolic syndrome. *Diabetes.* 2010;59:505–508.

Buchanan TA, Xiang AH. Gestational diabetes mellitus. *J Clin Invest.* 2005;115:485–491.

Burkhart DL, Sage J. Cellular mechanisms of tumour suppression by the retinoblastoma gene. *Nat Rev Cancer.* 2008;8:671–682.

Butler AA, Yakar S, Gewolb IH, Karas M, Okubo Y, LeRoith D. Insulin–like growth factor–I receptor signal transduction: At the interface between physiology and cell biology. *Comp Biochem Physiol B Biochem Mol Biol.* 1998;121:19–26.

Butler AE, Cao–Minh L, Galasso R, Rizza RA, Corradin A, Cobelli C, Butler PC. Adaptive changes in pancreatic beta cell fractional area and beta cell turnover in human pregnancy. *Diabetologia.* 2010;53:2167–76.

Butler AE, Janson J, Bonner–Weir S, Ritzel R, Rizza RA, Butler PC. β–Cell deficit and increased β–cell apoptosis in humans with type 2 diabetes. *Diabetes*. 2003;52:102–110.

Caban M, Owczarek K, Chojnacka K, Lewandowska U. Overview of polyphenols and polyphenol–rich extracts as modulators of IGF–1, IGF–1R, and IGFBP expression in cancer diseases. *J Funct Foods*. 2019;52:389–407.

Carr ME. Diabetes mellitus. A hypercoagulable state. *J Diabetes Complications*. 2001;15:44–45.

Cassidy FC, Charalambous M. Genomic imprinting, growth and maternal–fetal interactions. *J Exp Biol*. 2018;221(Suppl 1).pii: jeb164517.

Cetin I, de Santis MS, Taricco E, Radaelli T, Teng C, Ronzoni S, Spada E, Milani S, Pardi G. Maternal and fetal amino acid concentrations in normal pregnancies and in pregnancies with gestational diabetes mellitus. *Am J Obstet Gynecol*. 2005;192:610–617.

Chambers AF, Groom AC, MacDonald IC. Dissemination and growth of cancer cells in metastatic sites. *Nat Rev Cancer*. 2002;2:563–572.

Chen CL, Ip SM, Cheng D, Wong LC, Ngan HY. Loss of imprinting of the IGF–II and H19 genes in epithelial ovarian cancer. *Clin Cancer Res*. 2000;6:474–479.

Chen MY, Clark AJ, Chan DC, Ware JL, Holt SE, Chidambaram A, Fillmore HL, Broaddus WC. Wilms' tumor 1 silencing decreases the viability and chemoresistance of glioblastoma cells in vitro: A potential role for IGF–1R de–repression. *J Neurooncol*. 2011;103:87–102.

Chen Z, Liu J, Chu D, Shan Y, Ma G, Zhang H, Zhang XD, Wang P, Chen Q, Deng C, et al. A dual–specific IGF–I/II human engineered antibody domain inhibits IGF signaling in breast cancer cells. *Int J Biol Sci*. 2018;14:799–806.

Chon S, Choi MC, Lee YJ, Hwang YC, Jeong IK, Oh S, Ahn KJ, Chung HY, Woo JT, Kim SW, et al. Autoimmune hypoglycemia in a patient with characterization of insulin receptor autoantibodies. *Diabetes Metab J*. 2011:35:80–85.

Cianfarani S. Insulin–like growth factor–II: new roles for an old actor. *Front Endocrinol.* 2012;3:118.

Clemmons DR, Busby WH, Arai T, Nam TJ, Clarke JB, Jones JI, Ankrapp DK. Role of insulin–like growth factor binding proteins in the control of IGF actions. *Prog Growth Factor Res.* 1995;6:357–366.

Clemmons DR. Role of IGF–binding proteins in regulating IGF responses to changes in metabolism. *J Mol Endocrinol.* 2018;61:T139–T169.

Colomiere M, Permezel M, Riley C, Desoye G, Lappas M. Defective insulin signaling in placenta from pregnancies complicated by gestational diabetes mellitus. *Eur J Endocrinol.* 2009;160:567–578.

Contessa JN, Bhojani MS, Freeze HH, Rehemtulla A, Lawrence TS. Inhibition of n–linked glycosylation disrupts receptor tyrosine kinase signaling in tumor cells. *Cancer Res.* 2008;68:3803–3809.

Coolican SA, Samuel DS, Ewton DZ, McWade FJ, Florini JR. The mitogenic and myogenic actions of insulin–like growth factors utilize distinct signaling pathways. *J Biol Chem.* 1997;272:6653–6662.

Cruz PD, Hud JA. Excess insulin binding to insulin–like growth factor receptors: proposed mechanism for acanthosis nigricans. *J Invest Dermatol.* 1992;98:82S–85S.

Cubbon RM, Kearney MT, Wheatcroft SB. Endothelial IGF–1 Receptor signalling in diabetes and insulin resistance. *Trends Endocrinol Metab.* 2016;27:96–104.

Cui H, Cruz–Correa M, Giardiello FM, Hutcheon DF, Kafonek DR, Brandenburg S, Wu Y, He X, Powe NR, Feinberg AP. Loss of IGF2 imprinting: A potential marker of colorectal cancer risk. *Science.* 2003;299:1753–1755.

Czech MP, Massague J. Subunit structure and dynamics of the insulin receptor. *Fed Proc.* 1982;41:2719–2723.

De Rosa M, Pace U, Rega D, Costabile V, Duraturo F, Izzo P, Delrio P. Genetics, diagnosis and management of colorectal cancer. *Oncol Rep.* 2015;34:1087–1896.

de–Freitas–Junior JCM, Andrade–da–Costa J, Silva MC, Pinho SS. Glycans as regulatory elements of the insulin/IGF system: impact in cancer progression. *Int J Mol Sci.* 2017;18:1921.

Dennis JW, Laferte S. Tumor cell surface carbohydrate and the metastatic phenotype. *Cancer Metastasis Rev.* 1987;5:185–204.

Desoye G, Hartmann M, Blaschitz A, Dohr G, Hahn T, Kohnen G, Kaufmann P. Insulin receptors in syncytiotrophoblast and fetal endothelium of human placenta. Immunohistochemical evidence for developmental changes in distribution pattern. *Histochemistry.* 1994;101:277–285.

Di Minno MND, Lupoli R, Palmieri NM, Russolillo A, Buonauro A, Di Minno G. Aspirin resistance, platelet turnover, and diabetic angiopathy: A 2011 update. *Thromb Res.* 2012;129:341–344.

Diamant YZ, Metzger BE, Freinkel N, Shafrir E. Placental lipid and glycogen content in human and experimental diabetes mellitus. *Am J Obstet Gynecol.* 1982;144:5–11.

Diaz LE, Chuan YC, Lewitt M, Fernandez–Perez L, Carrasco–Rodríguez S, Sanchez–Gomez M, Flores–Morales A. IGF–II regulates metastatic properties of choriocarcinoma cells through the activation of the insulin receptor. *Mol Hum Reprod.* 2007;13:567–576.

Dimasuay KG, Boeuf P, Powell TL, Jansson T. Placental responses to changes in the maternal environment determine fetal growth. *Front Physiol.* 2016;7:12.

Dridi L, Seyrantepe V, Fougerat A, Pan X, Bonneil E, Thibault P, Moreau A, Mitchell GA, Heveker N, Cairo CW, et al. Positive regulation of insulin signaling by neuraminidase 1. *Diabetes.* 2013;62:2338–2346.

Duan C, Bauchat JR, Hsieh T. Phosphatidylinositol 3–kinase is required for insulin–like growth factor–I–induced vascular smooth muscle cell proliferation and migration. *Circ Res.* 2000;86:15–23.

Dubova EA, Pavlov KA, Lyapin VM, Kulikova GV, Shchyogolev AI, Sukhikh GT. Expression of insulin–like growth factors in the placenta in preeclampsia. *Bull Exp Biol Med.* 2014;157:103–107.

Dunn EJ, Philippou H, Ariëns RAS, Grant PJ. Molecular mechanisms involved in the resistance of fibrin to clot lysis by plasmin in subjects with type 2 diabetes mellitus. *Diabetologia.* 2006;49:1071–1080.

Dupont J, LeRoith D. Insulin and insulin–like growth factor I receptors: similarities and differences in signal transduction. *Horm Res.* 2001;55(Suppl 2):22–26.

Eades CE, Cameron DM, Evans JMM. Prevalence of gestational diabetes mellitus in Europe: A meta–analysis. *Diabetes Res Clin Pract.* 2017;129:173–181.

Elchalal U, Schaiff WT, Smith SD, Rimon E, Bildirici I, Nelson DM, Sadovsky Y. Insulin and fatty acids regulate the expression of the fat droplet–associated protein adipophilin in primary human trophoblasts. *Am J Obstet Gynecol.* 2005;193:1716–1723.

El–Shewy HM, Lee MH, Obeid LM, Jaffa AA, Luttrell LM. The insulin–like growth factor type 1 and insulin–like growth factor type 2/mannose–6–phosphate receptors independently regulate ERK1/2 activity in HEK293 cells. *J Biol Chem.* 2007;282:26150–26157.

Ericsson A, Hamark B, Jansson N, Johansson BR, Powell TL, Jansson T. Hormonal regulation of glucose and system A amino acid transport in first trimester placental villous fragments. *Am J Physiol Regul Integr Comp Physiol.* 2005;288:R656–662.

Escribano O, Beneit N, Rubio–Longás C, López–Pastor AR, Gómez–Hernández A. The role of insulin receptor isoforms in diabetes and its metabolic and vascular complications. *J Diabetes Res.* 2017;1403206.

Faienza MF, Santoro N, Lauciello R, Calabrò R, Giordani L, Di Salvo G, Ventura A, Delvecchio M, Perrone L, Del Giudice EM, et al. IGF2 gene variants and risk of hypertension in obese children and adolescents. *Pediatr Res.* 2010;67:340–344.

Fang J, Furesz TC, Lurent RS, Smith CH, Fant ME. Spatial polarization of insulin–like growth factor receptors on the human syncytiotrophoblast. *Pediatr Res.* 1997;41:258–265.

Federici M, Hribal ML, Ranalli M, Marselli L, Porzio O, Lauro D, Borboni P, Lauro R, Marchetti P, Melino G, et al. The common Arg972 polymorphism in insulin receptor substrate–1 causes apoptosis of human pancreatic islets. *FASEB J.* 2001;15:22–24.

Federici M, Lauro D, D'Adamo M, Giovannone B, Porzio O, Mellozzi M, Tamburrano G, Sbraccia P, Sesti G. Expression of insulin/IGF–I hybrid receptors is increased in skeletal muscle of patients with chronic primary hyperinsulinemia. *Diabetes.* 1998b;47:87–92.

Federici M, Porzio O, Lauro D, Borboni P, Giovannone B, Zucaro L, Hribal ML, Sesti G. Increased abundance of insulin/insulin–like growth factor–i hybrid receptors in skeletal muscle of obese subjects is correlated with *in vivo* insulin sensitivity. *J Clin Endocrinol Metab.* 1998a;83:2911–2915.

Federici M, Porzio O, Zucaro L, Giovannone B, Borboni P, Marini MA, Lauro D, Sesti G. Increased abundance of insulin/IGF–I hybrid receptors in adipose tissue from NIDDM patients. *Mol Cell Endocrinol.* 1997;135:41–47.

Federici M, Zucaro L, Porzio O, Massoud R, Borboni P, Lauro D, Sesti G. Increased expression of insulin/insulin–like growth factor–I hybrid receptors in skeletal muscle of noninsulin–dependent diabetes mellitus subjects. *J Clin Invest.* 1996;98:2887–2893.

Feldser D, Agani F, Iyer NV, Pak B, Ferreira G, Semenza GL. Reciprocal positive regulation of hypoxia–inducible factor 1alpha and insulin–like growth factor 2. *Cancer Res.* 1999;59:3915–3918.

Ferreira IG, Pucci M, Venturi G, Malagolini N, Chiricolo M, Dall'Olio F. Glycosylation as a main regulator of growth and death factor receptors signaling. *Int J Mol Sci.* 2018;19(2).pii: E580.

Ferreiro JL, Gómez–Hospital JA, Angiolillo DJ. Platelet abnormalities in diabetes mellitus. *Diab Vasc Dis Res.* 2010;7:251–259.

Fiaschi T, Chiarugi P. Oxidative stress, tumor microenvironment, and metabolic reprogramming: a diabolic liaison. *Int J Cell Biol.* 2012:762825.

Firth SM, Baxter RC. Cellular actions of the insulin–like growth factor binding proteins. *Endocr Rev.* 2002;23:824–854.

Firth SM, Baxter RC. Characterisation of recombinant glycosylation variants of insulin–like growth factor binding protein–3. *J Endocrinol.* 1999;160:379–387.

Flier JS, Kahn CR, Roth J, Bar RS. Antibodies that impair insulin receptor binding in an unusual diabetic syndrome with severe insulin resistance. *Science.* 1975;190:63–65.

Folkman J. Angiogenesis: An organizing principle for drug discovery? *Nat Rev Drug Discov*. 2007;6:273–286.

Forbes K, Shah VK, Siddals K, Gibson JM, Aplin JD, Westwood M. Statins inhibit insulin–like growth factor action in first trimester placenta by altering insulin–like growth factor1receptor glycosylation. *Mol Hum Rep*. 2015;2:105–114.

Forbes K, Westwood M, Baker PN, Aplin JD. Insulin–like growth factor I and II regulate the life cycle of trophoblast in the developing human placenta. *Am J Physiol Cell Physiol*. 2008;294:C1313–1322.

Forbes K, Westwood M. The IGF Axis and Placental Function. *Horm Res*. 2008;69:129–137.

Fowden AL, Sferruzzi–Perri AN, Coan PM, Constancia M, Burton GJ. Placental efficiency and adaptation: endocrine regulation. *J Physiol*. 2009;587(Pt 14):3459–3472.

Frasca F, Pandini G, Scalia P, Sciacca L, Mineo R, Costantino A, Goldfine ID, Belfiore A, Vigneri R. Insulin receptor isoform A, a newly recognized, high–affinity insulin–like growth factor II receptor in fetal and cancer cells. *Mol Cell Biol*. 1999;19:3278–3288.

Freier S, Weiss O, Eran M, Flyvbjerg A, Dahan R, Nephesh I, Safra T, Shiloni E, Raz I. Expression of the insulin–like growth factors and their receptors in adenocarcinoma of the colon. *Gut*. 1999;44:704–708.

Friedrich N, Thuesen B, Jørgensen T, Juul A, Spielhagen C, Wallaschofksi H, Linneberg A. The association between IGF–I and insulin resistance: a general population study in Danish adults. *Diabetes Care*. 2012;35:768–773.

Gatenby VK, Imrie H, Kearney M. The IGF–1 receptor and regulation of nitric oxide bioavailability and insulin signalling in the endothelium. *Eur J Physiol*. 2013;465:1065–1074.

Gaunt TR, Cooper JA, Miller GJ, Day IN, O'Dell SD. Positive associations between single nucleotide polymorphisms in the IGF2 gene region and body mass index in adult males. *Hum Mol Genet*. 2001;10:1491–1501.

Gebauer G, Jäger W, Lang N. mRNA expression of components of the insulin–like growth factor system in breast cancer cell lines, tissues, and metastatic breast cancer cells. *Anticancer Res.* 1998;18:1191–1195.

Girnita L, Worrall C, Takahashi S, Seregard S, Girnita A. Something old, something new and something borrowed: emerging paradigm of insulin–like growth factor type 1 receptor (IGF–1R) signaling regulation. *Cell Mol Life Sci.* 2014;71:2403–2427.

Gligorijević N, Penezić A, Nedić O. Influence of glyco–oxidation on complexes between fibrin(ogen) and insulin–like growth factor binding protein–1 in patients with diabetes mellitus type 2. *Free Radic Res.* 2017;51: 64–72.

Gligorijević N, Robajac D, Nedić O. An enhanced platelet sensitivity to IGF–I in patients with diabetes mellitus. *Biochem. (Mosc).* Article n press.

Graham ME, Kilby DM, Firth SM, Robinson PJ, Baxter RC. The *in vivo* phosphorylation and glycosylation of human insulin–like growth factor–binding protein–5. *Mol Cell Proteomics.* 2007;6:1392–1405.

Graves JA, Renfree MB. Marsupials in the age of genomics. *Annu Rev Genomics Hum Genet.* 2013;14:393–420.

Grill CJ, Sivaprasad U, Cohick WS. Constitutive expression of IGF–binding protein–3 by mammary epithelial cells alters signaling through Akt and p70S6 kinase. *J Mol Endocrinol.* 2002;29:153–162.

Grimberg A. P53 and IGFBP–3: Apoptosis and cancer protection. *Mol Genet Metab.* 2000;70:85–98.

Grivennikov SI, Greten FR, Karin M. Immunity, inflammation, and cancer. *Cell.* 2010;140:883–899.

Grulich–Henn J, Ritter J, Mesewinkel S, Heinrich U, Bettendorf M, Preissner KT. Transport of insulin–like growth factor–I across endothelial cell monolayers and its binding to the subendothelial matrix. *Exp Clin Endocrinol Diabetes.* 2002;110:67–73.

Gu D, O'Dell SD, Chen XH, Miller GJ, Day IN. Evidence of multiple causalsites affecting weight in the IGF2–INS–TH region of human chromosome 11. *Hum Genet.* 2002;110:173–181.

Gu J, Sato Y, Kariya Y, Isaji T, Taniguchi N, Fukuda T. A mutual regulation between cell–cell adhesion and N–glycosylation:

implication of the bisecting GlcNAc for biological functions. *J Proteome Res*. 2009;8:431–435.

Guzmán–Gutiérrez E, Arroyo P, Salsoso R, Fuenzalida B, Sáez T, Leiva A, Pardo F, Sobrevia L. Role of insulin and adenosine in the human placenta microvascular and macrovascular endothelial cell dysfunction in gestational diabetes mellitus. *Microcirculation*. 2014;21:26–37.

Haeusler RA, McGraw TE, Accili D. Biochemical and cellular properties of insulin receptor signalling. Nat Rev *Mol Cell Biol*. 2018;19:31–44.

Hakam A, Yeatman TJ, Lu L, Mora L, Marcet G, Nicosia SV, Karl RC, Coppola D. Expression of insulin–like growth factor–1 receptor in human colorectal cancer. *Hum Pathol*. 1999;30:1128–1133.

Hamilton GS, Lysiak JJ, Han VK, Lala PK. Autocrine–paracrine regulation of human trophoblast invasiveness by insulin–like growth factor (IGF)–II and IGF–binding protein (IGFBP)–1. *Exp Cell Res*. 1998;244:147–156.

Han VK, Bassett N, Walton J, Challis JR. The expression of insulin–like growth factor (IGF) and IGF–binding protein (IGFBP) genes in the human placenta and membranes: evidence for IGF–IGFBP interactions at the feto–maternal interface. *J Clin Endocrinol Metab*. 1996;81:2680–2693.

Hanahan D, Weinberg RA. Hallmarks of cancer: The next generation. *Cell*. 2011;144:646–674.

Hansen L, Hansen T, Clausen JO, Echwald SM, Urhammer SA, Rasmussen SK, Pedersen O. The Val985Met insulin–receptor variant in the Danish Caucasian population: Lack of associations with non–insulin–dependent diabetes mellitus or insulin resistance. *Am J Hum Genet*. 1997;60:1532–1535.

Hansen T, Bjørbaek C, Vestergaard H, Grønskov K, Bak JF, Pedersen O. Expression of insulin receptor spliced variants and their functional correlates in muscle from patients with non–insulin–dependent diabetes mellitus. *J Clin Endocrinol Metab*. 1993;77:1500–1505.

Harley CB, Kim NW, Prowse KR, Weinrich SL, Hirsch KS, West MD, Bacchetti S, Hirte HW, Counter CM, Greider CW. Telomerase, cell immortality, and cancer. *Cold Spring Harb Symp Quant Biol.* 1994;59:307–315.

Harris LK, Crocker IP, Baker PN, Aplin JD, Westwood M. IGF2 actions on trophoblast in human placenta are regulated by the insulin–like growth factor 2 receptor, which can function as both a signaling and clearance receptor. *Biol Reprod.* 2011;84:440–446.

Harris LK, Pantham P, Yong HEJ, Pratt A, Borg AJ, Crocker I, Westwood M, Aplin J, Kalionis B, Murthi P. The role of insulin–like growth factor 2 receptor–mediated homeobox gene expression in human placental apoptosis, and its implications in idiopathic fetal growth restriction. *Mol Hum Rep.* 2019;pii:gaz047.

Hart LM, Stolk RP, Dekker JM, Nijpels G, Grobbee DE, Heine RJ, Maassen A. Prevalence of variants in candidate genes for type 2 diabetes mellitus in the Netherlands: the Rotterdam study and the Hoorn study. *J Clin Endocrinol Metab.* 1999;84:1002–1006.

Haywood NJ, Slater TA, Matthews CJ, Wheatcroft SB. The insulin like growth factor and binding protein family: Novel therapeutic targets in obesity and diabetes. *Mol Metab.* 2019;19:86–96.

Heinemann l. Insulin assay standardization: leading to measures of insulin sensitivity and secretion for practical clinical care. *Diabetes Care.* 2010;33:e83.

Hers I. Insulin–like growth factor–1 potentiates platelet activation via the IRS–PI3Kα pathway. *Blood.* 2007;110:4243–4252.

Hiden U, Glitzner E, Hartmann M, Desoye G. Insulin and the IGF system in the human placenta of normal and diabetic pregnancies. *J Anat.* 2009;215:60–68.

Hiden U, Maier A, Bilban M, Ghaffari–Tabrizi N, Wadsack C, Lang I, Dohr G, Desoye G. Insulin control of placental gene expression shifts from mother to foetus over the course of pregnancy. *Diabetologia.* 2006;49:123–131.

Hills FA, Elder MG, Chard T, Sullivan MH. Regulation of human villous trophoblast by insulin–like growth factors and insulin–like growth factor–binding protein–1. *J Endocrinol.* 2004;183:487–496.

Hirschmugl B, Desoye G, Catalano P, Klymiuk I, Scharnagl H, Payr S, Kitzinger E, Schliefsteiner C, Lang U, et al. Maternal obesity modulates intracellular lipid turnover in the human term placenta. *Int J Obes (Lond)*. 2017;41:317–323.

Hoeck WG, Mukku VR. Identification of the major sites of phosphorylation in IGF binding protein 3. *J Cell Biochem*. 1994;56:262–273.

Holmes R, Porter H, Newcomb P, Holly JM, Soothill P. An immunohistochemical study of type I insulin–like growth factor receptors in the placentae of pregnancies with appropriately grown or growth restricted fetuses. *Placenta*. 1999;20:325–330.

Hsu PP, Sabatini DM. Cancer cell metabolism: Warburg and beyond. *Cell*. 2008;134:703–707.

Hu L, Chang L, Zhang Y, Zhai L, Zhang S, Qi Z, Yan H, Yan Y, Luo X, Zhang S, et al. Platelets express activated P2Y12 receptor in patients with diabetes mellitus. *Circulation*. 2017;136:817–833.

Hubbard SR, Till JH. Protein tyrosine kinase structure and function. *Annu Rev Biochem*. 2000;69:373–398.

Hunter RW, Hers I. Insulin/IGF–1 hybrid receptor expression on human platelets: consequences for the effect of insulin on platelet function. *J Thromb Haemost*. 2009;7:2123–2130.

Hwa V, Oh J, Rosenfeld RG. The insulin–like growth factor–binding protein (IGFBP) superfamily. *Endocr Rev*. 1999;20:761–787.

Hwang JB, Frost SC. Effect of alternative glycosylation on insulin receptor processing. *J Biol Chem*. 1999;274:22813–22820.

Igney FH, Krammer PH. Death and anti–death: tumour resistance to apoptosis. *Nat Rev Cancer*. 2002;2:277–288.

Imai Y, Clemmons DR. Roles of phosphatidylinositol 3–kinase and mitogen–activated protein kinase pathways in stimulation of vascular smooth muscle cell migration and deoxyriboncleic acid synthesis by insulin–like growth factor–I. *Endocrinology*. 1999;140:4228–4235.

Ingermann AR, Yang YF, Han J, Mikami A, Garza AE, Mohanraj L, Fan L, Idowu M, Ware JL, Kim HS, et al. Identification of a novel cell death receptor mediating IGFBP–3–induced anti–tumor effects in breast and prostate cancer. *J Biol Chem*. 2010;285:30233–30246.

Iñiguez G, Castro JJ, Garcia M, Kakarieka E, Johnson MC, Cassorla F, Mericq V. IGF–IR signal transduction protein content and its activation by IGF–I in human placentas: relationship with gestational age and birth weight. *PLoS One*. 2014;9:e102252.

Iñiguez G, Gallardo P, Castro JJ, Gonzalez R, Garcia M, Kakarieka E, San Martin S, Johnson MC, Mericq V, Cassorla F. Klotho gene and protein in human placentas according to birth weight and gestational age. *Front Endocrinol (Lausanne)*. 2019;9:797.

International Diabetes Federation, *IDF Diabetes Atlas*, International Diabetes Federation, Brussels, Belgium, 8th edition, 2017, http://www.diabetesatlas.org.

Irwin JC, Suen LF, Faessen GH, Popovici RM, Giudice LC. Insulin–like growth factor (IGF)–II inhibition of endometrial stromal cell tissue inhibitor of metalloproteinase–3 and IGF–binding protein–1 suggests paracrine interactions at the decidua: trophoblast interface during human implantation. *J Clin Endocrinol Metab*. 2001;86:2060–2064.

Itkonen HM, Mills IG. N–linked glycosylation supports cross–talk between receptor tyrosine kinases and androgen receptor. *PLoS One*. 2013;8:e65016.

Jackson SP, Bartek J. The DNA–damage response in human biology and disease. *Nature*. 2009;461:1071–1078.

Jafari E, Gheysarzadeh A, Mahnam K, Shahmohammadi R, Ansari A, Bakhtyari H, Mofid MR. *In silico* interaction of insulin–like growth factor binding protein 3 with insulin–like growth factor 1. *Res Pharmaceut Sci*. 2018;13:332–342.

Jafri MA, Ansari SA, Alqahtani MH, Shay JW. Roles of telomeres and telomerase in cancer, and advances in telomerase–targeted therapies. *Genome Med*. 2016;8:69.

Janssen JAMJL. IGF–I and the endocrinology of aging. *Curr Opin Endocr Metab Res*. 2019;5:1–6.

Jansson N, Rosario FJ, Gaccioli F, Lager S, Jones HN, Roos S, Jansson T, Powell TL. Activation of placental mTOR signaling and amino acid transporters in obese women giving birth to large babies. *J Clin Endocrinol Metab*. 2013;98:105–113.

Jansson T, Aye IL, Goberdhan DC. The emerging role of mTORC1 signaling in placental nutrient–sensing. *Placenta*. 2012;33(Suppl 2):e23–e29.

Jansson T, Powell TL. Human placental transport in altered fetal growth: does the placenta function as a nutrient sensor? *Placenta*. 2006;27:91–97.

Jeong KH, Oh SJ, Chon S, Lee MH. Generalized acanthosis nigricans related to type B insulin resistance syndrome: A case report. *Cutis*. 2010;86:299–302.

Jiang BH, Liu LZ. PI3K/PTEN signaling in angiogenesis and tumorigenesis. *Adv Cancer Res*. 2009;102:19–65.

Jiang H, Xun P, Luo G, Wang Q, Cai Y, Zhang Y, Yu B. Levels of insulin–like growth factors and their receptors in placenta in relation to macrosomia. *Asia Pac J Clin Nutr*. 2009;18:171–178.

Jirkovská M, Kubínová L, Janácek J, Moravcová M, Krejcí V, Karen P. Topological properties and spatial organization of villous capillaries in normal and diabetic placentas. *J Vasc Res*. 2002;39:268–278.

Jones JI, Busby WH Jr, Wright G, Smith CE, Kimack NM, Clemmons DR. Identification of the sites of phosphorylation in insulin–like growth factor binding protein–1. Regulation of its affinity by phosphorylation of serine 101. *J Biol Chem*. 1993;268:1125–1131.

Jones R. Cytotoxic chemotherapy: Clinical aspects. *Medicine*. 2016;44:25–29.

Juul A. Serum levels of insulin–like growth factor I and its binding proteins in health and disease. *Growth Horm IGF Res*. 2003;13:113–170.

Kadakia R, Josefson J. The relationship of insulin–like growth factor 2 to fetal growth and adiposity. *Horm Res Paediatr*. 2016;85:75–82.

Kakouros N, Rade JJ, Kourliouros A, Resar JR. Platelet function in patients with diabetes mellitus: From a theoretical to a practical perspective. *Int J Endocrinol*. 2011:742719.

Kanasaki K, Kalluri R. The biology of preeclampsia. *Kidney Int.* 2009;76:831–837.

Kaneda A, Wang CJ, Cheong R, Timp W, Onyango P, Wen B, Iacobuzio–Donahue CA, Ohlsson R, Andraos R, Pearson MA, et al. Enhanced sensitivity to IGF–II signaling links loss of imprinting of IGF2 to increased cell proliferation and tumor risk. *Proc Natl Acad Sci USA.* 2007;104:20926–20931.

Kaplan S, Cohen P. The somatomedin hypothesis 2007: 50 years later. *J Cin Endocrinol Metab.* 2007;92:4529–4535.

Karolczak–Bayatti M, Forbes K, Horn J, Teesalu T, Harris LK, Westwood M, Aplin JD. IGF signalling and endocytosis in the human villous placenta in early pregnancy as revealed by comparing quantum dot conjugates with a soluble ligand. *Nanoscale.* 2019;11:12285–12295.

Kataoka H, Tanaka H, Nagaike K, Uchiyama S, Itoh H. Role of cancer cell–stroma interaction in invasive growth of cancer cells. *Hum Cell.* 2003;16:1–14.

Katayama Y, Uchino J, Chihara Y, Tamiya N, Kaneko Y, Yamada T, Takayama K. Tumor Neovascularization and Developments in Therapeutics. *Cancers (Basel).* 2019;11:316.

Kellerer M, Sesti G, Seffer E, Obermaier–Kusser B, Pongratz DE, Mosthaf L, Häring HU. Altered pattern of insulin receptor isotypes in skeletal muscle membranes of Type 2 (non–insulin–dependent) diabetic subjects. *Diabetologia.* 1993;36:628–632.

Kessenbrock K, Plaks V, Werb Z. Matrix metalloproteinases: regulators of the tumor microenvironment. *Cell.* 2010;141:52–67.

Kharb S, Nanda S. Patterns of biomarkers in cord blood during pregnancy and preeclampsia. *Curr Hypertens Rev.* 2017;13:57–64.

Kharb S, Panjeta P, Ghalaut VS, Bala J, Nanda S. Biomarkers in preeclamptic women with normoglycemia and hyperglycemia. *Curr Hypertens Rev.* 2016;12:228–233.

Kim JG, Kang MJ, Yoon YK, Kim HP, Park J, Song SH, Han SW, Park JW, Kang GH, Kang KW, et al. Heterodimerization of glycosylated insulin–like growth factor–1 receptors and insulin receptors in cancer cells sensitive to anti–IGF1R antibody. *PLoS One.* 2012;7:e33322.

Kim JH, Bae HY, Kim SY. Clinical marker of platelet hyperreactivity in diabetes mellitus. *Diabetes Metab J*. 2013;37:423–428.

Kim KW, Bae SK, Lee OH, Bae MH, Lee MJ, Park BC. Insulin–like growth factor II induced by hypoxia may contribute to angiogenesis of human hepatocellular carcinoma. *Cancer Res*. 1998;58:348–351.

Kim S, Garcia A, Jackson SP, Kunapuli SP. Insulin–like growth factor–1 regulates platelet activation through PI3–Kα isoform. *Blood*. 2007;110:4206–4213.

King GL, Kahn CR, Rechler MM, Nissley SP. Direct demonstration of separate receptors for growth and metabolic activities of insulin and multiplication–stimulating activity (an insulin–like growth factor) using antibodies to the insulin receptor. *J Clin Invest*. 1980;66:130–140.

Kingdom J, Huppertz B, Seaward G, Kaufmann P. Development of the placental villous tree and its consequences for fetal growth. *Eur J Obstet Gynecol Reprod Biol*. 2000;92:35–43.

Klaver E, Zhao P, May M, Flanagan–Steet H, Freeze HH, Gilmore R, Wells L, Contessa J, Steet R. Selective inhibition of N–linked glycosylation impairs receptor tyrosine kinase processing. *Dis Models Mechan*. 2019;12:pii:dmm039602.

Knofler M, Sooranna SR, Daoud G, Whitley GS, Markert UR, Xia Y, Cantiello H, Hauguel–de–Mouzon S. Trophoblast signalling: Knowns and unknowns – a workshop report. *Placenta*. 2005;26(Suppl A):S49–S51.

Kolch W. Meaningful relationships: The regulation of the Ras/Raf/MEK/ERK pathway by protein interactions. *Biochem J*. 2000;351Pt2:289–305.

Kornfeld S. Structure and function of the mannose 6–phosphate/insulinlike growth factor II receptors. *Annu Rev Biochem*. 1992;61:307–330.

Krakhmal NV, Zavyalova MV, Denisov EV, Vtorushin SV, Perelmuter VM. Cancer invasion: Patterns and mechanisms. *Acta Naturae*. 2015;7:17–28.

Kruis T, Klammt J, Galli–Tsinopoulou A, Wallborn T, Schlicke M, Müller E, Kratzsch J, Körner A, Odeh R, Kiess W, et al. Heterozygous mutation within a kinase–conserved motif of the insulin–like growth factor I receptor causes intrauterine and postnatal growth retardation. *J Clin Endocrinol Metab*. 2010;95:1137–1142.

Kühnl A, Kaiser M, Neumann M, Fransecky L, Heesch S, Radmacher M, Marcucci G, Bloomfield CD, Hofmann WK, Thiel E, et al. High expression of IGFBP2 is associated with chemoresistance in adult acute myeloid leukemia. *Leuk Res*. 2011;35:1585–1590.

Kurien BT, Hensley K, Bachmann M, Scofield RH. Oxidatively modified autoantigens in autoimmune diseases. *Free Radic Biol Med*. 2006;41:549–556.

Lacey H, Haigh T, Westwood M, Aplin JD. Mesenchymally–derived insulin–like growth factor 1 provides a paracrine stimulus for trophoblast migration. *BMC Dev Biol*. 2002;2:5.

Lane DP. Cancer. p53, guardian of the genome. *Nature*. 1992;358:15–16.

Langer CJ, Novello S, Park K, Krzakowski M, Karp DD, Mok T, Benner RJ, Scranton JR, Olszanski AJ, Jassem J. Randomized, phase III trial of first–line figitumumab in combination with paclitaxel and carboplatin versus paclitaxel and carboplatin alone in patients with advanced non–small–cell lung cancer. *J Clin Oncol*. 2014;32:2059–2066.

Lassance L, Miedl H, Absenger M, Diaz–Perez F, Lang U, Desoye G, Hiden U. Hyperinsulinemia stimulates angiogenesis of human fetoplacental endothelial cells: A possible role of insulin in placental hypervascularization in diabetes mellitus. *J Clin Endocrinol Metab*. 2013;98:E1438–1447.

Laviola L, Perrini S, Belsanti G, Natalicchio A, Montrone C, Leonardini A, Vimercati A, Scioscia M, Selvaggi L, Giorgino R, et al. Intrauterine growth restriction in humans is associated with abnormalities in placental insulin–like growth factor signaling. *Endocrinology*. 2005;146:1498–1505.

Laybutt DR, Kaneto H, Hasenkamp W, Grey S, Jonas JC, Sgroi DC, Groff A, Ferran C, Bonner–Weir S, Sharma A, et al. Increased expression of antioxidant and antiapoptotic genes in islets that may contribute to β–

cell survival during chronic hyperglycemia. *Diabetes*. 2002;51:413–423.

Le Roith D, Bondy C, Yakar S, Liu JL, Butler A. The somatomedin hypothesis: 2001. *Endocr Rev*. 2001;22:53–74.

Le Roith D. The insulin–like growth factor system. *Exp Diabesity Res*. 2003;4:205–212.

Lee H, Jang HC, Park HK, Metzger BE, Cho NH. Prevalence of type 2 diabetes among women with a previous history of gestational diabetes mellitus. *Diabetes Res Clin Pract*. 2008;81:124–129.

Lichtor T, Kurpakus MA, Gurney ME. Expression of insulin–like growth factors and their receptors in human meningiomas. *J Neurooncol*. 1993;17:183–190.

Liu B, Xu Y, Voss C, Qiu FH, Zhao MZ, Liu YD, Nie J, Wang ZL. Altered protein expression in gestational diabetes mellitus placentas provides insight into insulin resistance and coagulation/fibrinolysis pathways. *PLoS One*. 2012;7:e44701.

Liu D, Zhang X, Gao J, Palombo M, Gao D, Chen P, Sinko PJ. Core functional sequence of C–terminal GAG–binding domain directs cellular uptake of IGFBP–3–derived peptides. *Protein Pept Lett*. 2014;21:124–131.

Livingstone C. IGF2 and cancer. *Endocr Relat Cancer*. 2013;20:R321–339.

Lobel P, Dahms NM, Breitmeyer J, Chirgwin JM, Kornfeld S. Cloning of the bovine 215–kDa cation–independent mannose 6–phosphate receptor. *Proc Natl Acad Sci USA*. 1987;84:2233–2237.

Lobel P, Dahms NM, Kornfeld S. Cloning and sequence analysis of the cation–independent mannose 6–phosphate receptor. *J Biol Chem*. 1988;263:2563–2570.

Long L, Navab R, Brodt P. Regulation of the Mr 72,000 type IV collagenase by the type I insulin–like growth factor receptor. *Cancer Res*. 1998;58:3243–3247.

Longo N, Langley SD, Griffin LD, Elsas LJ. Two mutations in the insulin receptor gene of a patient with Leprechaunism: Application to prenatal diagnosis. *J Clin Endocrinol Metab*. 1995;80:1496–1501.

Longo N, Wang Y, Pasquali M. Progressive decline in insulin levels in Rabson–Mendenhall syndrome. *J Clin Endocrinol Metab*. 1999;84:2623–2629.

Longo N, Wang Y, Smith SA, Langley SD, DiMeglio LA, Giannella–Neto D. Genotype–phenotyte correlation in inherited severe insulin resistance. *Hum Mol Genet*. 2002;11:1465–1475.

Lou M, Garrett TPJ, McKern NM, Hoyne PA, Epa VC, Bentley JD, Lovrecz GO, Cosgrove LJ, Frenkel MJ, Ward CW. The first three domains of the insulin receptor differ structurally from the insulin–like growth factor 1 receptor in the regions governing ligand specificity. *Proc Natl Acad Sci USA*. 2006;103:12429–12434.

Loukovaara M, Leinonen P, Teramo K, Nurminen E, Andersson S, Rutanen EM. Effect of maternal diabetes on phosphorylation of insulin–like growth factor binding protein–1 in cord serum. *Diab Med*. 2005;22:434–439.

Ma M, Zhou QJ, Xiong Y, Li B, Li XT. Preeclampsia is associated with hypermethylation of IGF–1 promoter mediated by DNMT1. *Am J Transl Res*. 2018;10:16–39.

Maiza JC, Caron–Debarle M, Vigouroux C, Schneebeli S. Anti–insulin receptor antibodies related to hypoglycemia in a previously diabetic patient. *Diabetes Care*. 2013;36:e77.

Malaguarnera R, Belfiore A. The insulin receptor: a new target for cancer therapy. *Front Endocrinol (Lausanne)*. 2011;2:93.

Malek R, Chong AY, Lupsa BC, Lungu AO, Cochran EK, Soos MA, Semple RK, Balow JE, Gorden P. Treatment of type B insulin resistance: A novel approach to reduce insulin receptor autoantibodies. *J Clin Endocrinol Metab*. 2010;95:3641–3647.

Man YG, Stojadinovic A, Mason J, Avital I, Bilchik A, Bruecher B, Protic M, Nissan A, Izadjoo M, Zhang X, et al. Tumor–infiltrating immune cells promoting tumor invasion and metastasis: Existing theories. *J Cancer*. 2013;4:84–95.

Mani SA, Guo W, Liao MJ, Eaton EN, Ayyanan A, Zhou AY, Brooks M, Reinhard F, Zhang CC, Shipitsin M, et al. The epithelial–mesenchymal transition generates cells with properties of stem cells. *Cell.* 2008;133:704–715.

Martino J, Sebert S, Segura MT, García–Valdés L, Florido J, Padilla MC, Marcos A, Rueda R, McArdle HJ, Budge H, et al. Maternal body weight and gestational diabetes differentially influence placental and pregnancy outcomes. *J Clin Endocrinol Metab.* 2016;101:59–68.

Maruo T, Murata K, Matsuo H, Samoto T, Mochizuki M. Insulin–like growth factor–I as a local regulator of proliferation and differentiated function of the human trophoblast in early pregnancy. *Early Pregnancy.* 1995;1:54–61.

Mayama R, Izawa T, Sakai K, Suciu N, Iwashita M. Improvement of insulin sensitivity promotes extravillous trophoblast cell migration stimulated by insulin–like growth factor–I. *Endocr J.* 2013;60:359–368.

McKinnon T, Chakraborty C, Gleeson LM, Chidiac P, Lala PK. Stimulation of human extravillous trophoblast migration by IGF–II is mediated by IGF type 2 receptor involving inhibitory G protein(s) and phosphorylation of MAPK. *J Clin Endocrinol Metab.* 2001;86:3665–3674.

Migdalis IN, Kalogeropoulou K, Kalantzis L, Nounopoulos C, Bouloukos A, Samartzis M. Insulin–like growth factor I and IGF–I receptors in diabetic patients with neuropathy. *Diabetic Med.* 1995;12:823–827.

Milio LA, Hu J, Douglas GC. Binding of insulin–like growth factor I to human trophoblast cells during differentiation in vitro. *Placenta.* 1994;15:641–651.

Mira E, Mañes S, Lacalle RA, Márquez G, Martínez–A C. Insulin–like growth factor I–triggered cell migration and invasion are mediated by matrix metalloproteinase–9. *Endocrinology.* 1999;140:1657–1664.

Modestino AE, Skowronski EA, Pruitt C, Taub PR, Herbst K, Schmid–Schönbein GW, Heller MJ, Mills PJ. Elevated resting and postprandial digestive proteolytic activity in peripheral blood of individuals with type–2 diabetes mellitus, with uncontrolled cleavage of insulin receptors. *J Am Coll Nutr.* 2019;9:1–8.

Mol BWJ, Roberts CT, Thangaratinam S, Magee LA, de Groot CJM, Hofmeyr GJ. Pre–eclampsia. *Lancet*. 2016;387:999–1011.

Montazami N, Aghapour M, Farajnia S, Baradaran B. New insights into the mechanisms of multidrug resistance in cancers. *Cell Mol Biol (Noisy–le–grand)*. 2015;61:70–80.

Mooi WJ, Peeper DS. Oncogene–induced cell senescence – halting on the road to cancer. *N Engl J Med*. 2006;355:1037–1046.

Moore SF, Williams CM, Brown E, Blair TA, Harper MT, Coward RJ, Poole AW, Hers I. Loss of the insulin receptor in murine megakaryocytes/platelets causes thrombocytosis and alterations in IGF signaling. *Cardiovasc Res*. 2015;107:9–19.

Moorehead RA, Hojilla CV, De Belle I, Wood GA, Fata JE, Adamson ED, Watson KL, Edwards DR, Khokha R. Insulin–like growth factor–II regulates PTEN expression in the mammary gland. *J Biol Chem*. 2003;278:50422–50427.

Moses AC, Young SC, Morrow LA, O'Brien M, Clemmons DR. Recombinant human insulin–like growth factor I increases insulin sensitivity and improves glycemic control in type II diabetes. *Diabetes*. 1996;45:91–100.

Mosthaf L, Eriksson J, Häring HU, Groop L, Widen E, Ullrich A. Insulin receptor isotype expression correlates with risk of non–insulin–dependent diabetes. *Proc Natl Acad Sci USA*. 1993;90:2633–2635.

Mosthaf L, Vogt B, Häring HU, Ullrich A. Altered expression of insulin receptor types A and B in the skeletal muscle of non–insulin–dependent diabetes mellitus patients. *Proc Natl Acad Sci USA*. 1991;88:4728–4730.

Motyka B, Korbutt G, Pinkoski MJ, Heibein JA, Caputo A, Hobman M, Barry M, Shostak I, Sawchuk T, Holmes CF, et al. Mannose 6–phosphate/insulin–like growth factor II receptor is a death receptor for granzyme B during cytotoxic T cell–induced apoptosis. *Cell*. 2000;103:491–500.

Mrizak I, Grissa O, Henault B, Fekih M, Bouslema A, Boumaiza I, Zaouali M, Tabka Z, Khan NA. Placental infiltration of inflammatory markers in gestational diabetic women. *Gen Physiol Biophys*. 2014;33:169–176.

Nadimpalli SK, Amancha PK. Evolution of mannose 6–phosphate receptors (MPR300 and 46): Lysosomal enzyme sorting proteins. *Curr Protein Pept Sci*. 2010;11:68–90.

Nedić O, Robajac D, Šunderić M, Miljuš G, Đukanović B, Malenković V. Detection and identification of the oxidized insulin–like growth factor binding proteins and receptors in patients with colorectal carcinoma. *Free Rad Biol Med*. 2013;65:1195–1200.

Negrini S, Gorgoulis VG, Halazonetis TD. Genomic instability – an evolving hallmark of cancer. Nat Rev *Mol Cell Biol*. 2010;11:220–228.

Neumann GM, Marinaro JA, Bach LA. Identification of N–glycosylation sites and partial characterization of carbohydrate structure and disulfide linkages of human insulin–like growth factor binding protein 6. *Biochemistry*. 1998;37:6572–6585.

Nimptsch K, Konigorski S, Pischon T. Diagnosis of obesity and use of obesity biomarkers in science and clinical medicine. *Metab Clin Exp*. 2019;92:61–70.

Norgren S, Zierath J, Galuska D, Wallberg–Henriksson H, Luthman H. Differences in the ratio of RNA encoding two isoforms of the insulin receptor between control and NIDDM Patients. The RNA variant without exon 11 predominates in both groups. *Diabetes*. 1993;42:675–681.

Novosyadlyy R, Le Roith D. Insulin–like growth factors and insulin: at the crossroad between tumor development and longevity. *J Gerontol A Bio Sci Med Sci*. 2012;67:640–651.

O'Connor KG, Tobin JD, Harman SM, Plato CC, Roy TA, Sherman SS, Blackman MR. Serum levels of insulin–like growth factor–I are related to age and not to body composition in healthy women and man. *J Gerontol Med Sci*. 1988;53A:M176–M182.

Ogawa O, Becroft DM, Morison IM, Eccles MR, Skeen JE, Mauger DC, Reeve AE. Constitutional relaxation of insulin–like growth factor II gene imprinting associated with Wilms' tumour and gigantism. *Nat Genet*. 1993;5:408–412.

O'Gorman DB, Weiss J, Hettiaratchi A, Firth SM, Scott CD. Insulin–like growth factor–II/mannose 6–phosphate receptor overexpression reduces growth of choriocarcinoma cells in vitro and in vivo. *Endocrinology*. 2002;143:4287–4294.

Ohtsubo K, Marth JD. Glycosylation in cellular mechanisms of health and disease. *Cell*. 2006;126:855–867.

Olson LJ, Castonguay AC, Lasanajak Y, Peterson FC, Cummings RD, Smith DF, Dahms NM. Identification of a fourth mannose 6–phosphate binding site in the cation–independent mannose 6–phosphate receptor. *Glycobiology*. 2014;25:591–606.

Ong K, Kratzsch J, Kiess W, Costello M, Scott C, Dunger D. Size at birth and cord blood levels of insulin, insulin–like growth factor I (IGF–I), IGF–II, IGF–binding protein–1 (IGFBP–1), IGFBP–3, and the soluble IGF–II/mannose–6–phosphate receptor in term human infants. The ALSPAC Study Team. Avon Longitudinal Study of Pregnancy and Childhood. *J Clin Endocrinol Metab*. 2000;85:4266–4269.

Orme SM, McNally RJ, Cartwright RA, Belchetz PE. Mortality and cancer incidence in acromegaly: A retrospective cohort study. United Kingdom Acromegaly Study Group. *J Clin Endocrinol Metab*. 1998;83:2730–2734.

Pathmaperuma AN, Mana P, Cheung SN, Kugathas K, Josiah A, Koina ME, Broomfield A, Delghingaro–Augusto V, Ellwood DA, Dahlstrom JE, et al. Fatty acids alter glycerolipid metabolism and induce lipid droplet formation, syncytialisation and cytokine production in human trophoblasts with minimal glucose effect or interaction. *Placenta*. 2010;31:230–239.

Peng HY, Xue M, Xia AB. Study on changes of IGF–I and leptin levels in serum and placental tissue of preeclampsia patients and their associativity. *Xi Bao Yu Fen Zi Mian Yi Xue Za Zhi*. 2011;27:192–194.

Perkins E, Murphy SK, Murtha AP, Schildkraut J, Jirtle RL, Demark–Wahnefried W, Forman MR, Kurtzberg J, Overcash F, Huang Z, et al. Insulin–like growth factor 2/H19 methylation at birth and risk of overweight and obesity in children. *J Pediatr*. 2012;161:31–39.

Phiske MM. An approach to acanthosis nigricans. *Indian Dermatol Online J.* 2014;5:239–249.

Pollak M. The insulin receptor/insulin–like growth factor receptor family as a therapeutic target in oncology. *Clin Cancer Res.* 2012;18:40–50.

Pomero F, Di Minno MND, Fenoglio L, Gianni M, Ageno W, Dentali F. Is diabetes a hypercoagulable state? A critical appraisal. *Acta Diabetol.* 2015;52:1007–1016.

Porzio O, Federici M, Hribal ML, Lauro D, Accili D, Lauro R, Borboni P, Sesti G. The $Gly^{972} \rightarrow Arg$ amino acid polymorphism in IRS–1 impairs insulin secretion in pancreatic β cells. *J Clin Invest.* 1999;104:357–364.

Pshezhetsky AV, Ashmarina LI. Desialylation of surface receptors as a new dimension in cell signaling. *Biochemistry (Moscow).* 2013;78:736–745.

Rabson SM, Mendenhall EN. Familial hypertrophy of pineal body, hyperplasia of adrenal cortex and diabetes mellitus. *Am J Clin Pathol.* 1956;26:283–290.

Radaelli T, Varastehpour A, Catalano P, Hauguel–de–Mouzon S. Gestational diabetes induces placental genes for chronic stress and inflammatory pathways. *Diabetes.* 2003;52:2951–2958.

Randriamboavonjy V, Fleming I. Insulin, insulin resistance, and platelet signaling in diabetes. *Diabetes Care.* 2009;32:528–530.

Renehan AG, Zwahlen M, Minder C, O'Dwyer ST, Shalet SM, Egger M. Insulin–like growth factor (IGF)–I, IGF binding protein–3, and cancer risk: Systematic review and meta–regression analysis. *Lancet.* 2004;363:1346–1453.

Reuter S, Gupta SC, Chaturvedi MM, Aggarwal BB. Oxidative stress, inflammation, and cancer: How are they linked? *Free Radic Biol Med.* 2010;49:1603–1616.

Ribatti D. A revisited concept: Contact inhibition of growth. From cell biology to malignancy. *Exp Cell Res.* 2017a;359:17–19.

Ribatti D. Endogenous inhibitors of angiogenesis: a historical review. *Leuk Res.* 2009;33:638–644.

Ribatti D. The concept of immune surveillance against tumors. The first theories. *Oncotarget*. 2017b;8:7175–7180.

Rinderknecht E, Humbel RE. The amino acid sequence of human insulin–like growth factor I and its structural homology with proinsulin. *J Biol Chem*. 1978;253:2769–2776.

Robajac D, Križáková M, Masnikosa R, Miljuš G, Šunderić M, Nedić O, Katrlík J. Sensitive glycoprofiling of insulin–like growth factor receptors isolated from colon tissue of patients with colorectal carcinoma using lectin-based protein microarray. *Int J Biol Macromol*. 2019 (under review).

Robajac D, Masnikosa R, Filimonović D, Miković Ž, Nedić O. N–glycosylation pattern of human placental insulin–like growth factor and insulin receptors in well–controlled pregestational diabetes mellitus. *J Med Biochem*. 2012;31:205–210.

Robajac D, Masnikosa R, Miković Ž, Mandić V, Nedić O. Oxidation of placental insulin and insulin–like growth factor receptors in mothers with diabetes mellitus or preeclampsia complicated with intrauterine growth restriction. *Free Radic Res*. 2015;49:984–989.

Robajac D, Masnikosa R, Miković Ž, Nedić O. Gestation–associated changes in the glycosylation of placental insulin and insulin–like growth factor receptors. *Placenta*. 2016a;39:70–76.

Robajac D, Masnikosa R, Vanhooren V, Libert C, Miković Ž, Nedić O. The N–glycan profile of placental membrane glycoproteins alters during gestation and ageing. *Mech Ageing Dev*. 2014;138:1–9.

Robajac D, Vanhooren V, Masnikosa R, Miković Ž, Mandić V, Libert C, Nedić O. Preeclampsia transforms membrane N–glycome in human placenta. *Exp Mol Pathol*. 2016b;100:26–30.

Robajac D, Zámorová M, Katrlík J, Miković Ž, Nedić O. Screening for the best detergent for the isolation of placental membrane proteins. *Int J Biol Macromol*. 2017;102:431–437.

Rogers J, Wiltrout L, Nanu L, Fant ME. Developmentally regulated expression of IGF binding protein–3 (IGFBP–3) in human placental fibroblasts: Effect of exogenous IGFBP3 on IGF–1 action. *Regul Pept*. 1996;61:189–195.

Romano G. The complex biology of the receptor for the insulin–like growth factor–1. *Drug News Perspect*. 2003;16:525–531.

Roos S, Lagerlof O, Wennergren M, Powell TL, Jansson T. Regulation of amino acid transporters by glucose and growth factors in cultured primary human trophoblast cells is mediated by mTOR signaling. *Am J Physiol Cell Physiol*. 2009;297:C723–C731.

Ruiz–Palacios M, Prieto–Sánchez M, Ruiz–Alcaraz A, Blanco–Carnero J, Sanchez–Campillo M, Parrilla J, Larqué E. Insulin treatment may enhance fatty acid carriers in placentas from gestational diabetes subjects. *Int J Mol Sci*. 2017b;6;18(6).

Ruiz–Palacios M, Ruiz–Alcaraz AJ, Sanchez–Campillo M, Larqué E. Role of insulin in placental transport of nutrients in gestational diabetes mellitus. *Ann Nutr Metab*. 2017;70:16–25.

Russo VC, Bach LA, Fosang AJ, Baker NL, Werther GA. Insulin like growth factor binding protein–2 binds to cell surface proteoglycans in the rat brain olfactory bulb. *Endocrinology*. 1997;138:4858–4856.

Ryan PD, Goss PE. The emerging role of the insulin–like growth factor pathway as a therapeutic target in cancer. *Oncologist*. 2008;13:16–24.

Saito–Hakoda A, Nishii A, Uchida T, Kikuchi A, Kanno J, Fujiwara I, Kure S. A follow–up during puberty in a Japanese girl with type A insulin resistance due to a novel mutation in *INSR*. *Clin Pediatr Endocrinol*. 2018;27:53–57.

Sakai K, Lowman HB, Clemmons DR. Increases in free, unbound insulin–like growth factor I enhance insulin responsiveness in human hepatoma G2 cells in culture. *J Biol Chem*. 2002;277:13620–13627.

Samani AA, Yakar S, LeRoith D, Brodt P. The role of the IGF system in cancer growth and metastasis: Overview and recent insights. *Endocr Rev*. 2007;28:20–47.

Sandberg AC, Engberg C, Lake M, von Holst H, Sara VR. The expression of insulin–like growth factor I and insulin–like growth factor II genes in the human fetal and adult brain and in glioma. *Neurosci Lett*. 1988;93:114–119.

Sandovici I, Hoelle K, Angiolini E, Constância M. Placental adaptations to the maternal–fetal environment: Implications for fetal growth and developmental programming. *Reprod Biomed Online*. 2012;25:68–89.

Sarfstein R, Pasmanik–Chor M, Yeheskel A, Edry L, Shomron N, Warman N, Wertheimer E, Maor S, Shochat L, Werner H. Insulin–like growth factor–I receptor (IGF–IR) translocates to nucleus and autoregulates IGF–IR gene expression in breast cancer cells. *J Biol Chem*. 2012;287:2766–2776.

Sasaoka T, Ishiki M, Wada T, Hori H, Hirai H, Haruta T, Ishihara H, Kobayashi M. Tyrosine phosphorylation–dependent and –independent role of Shc in the regulation of IGF–1–induced mitogenesis and glycogen synthesis. *Endocrinology*. 2001;142:5226–5235.

Sayer RA, Lancaster JM, Pittman J, Gray J, Whitaker R, Marks JR, Berchuck A. High insulin–like growth factor–2 (IGF–2) gene expression is an independent predictor of poor survival for patients with advanced stage serous epithelial ovarian cancer. *Gynecol Oncol*. 2005;96:355–361.

Sbraccia P, D'Adamo M, Leonetti F, Caiola S, Iozzo P, Giaccari A, Buongiorno A, Tamburrano G. Chronic primary hyperinsulinaemia is associated with altered insulin receptor mRNA splicing in muscle of patients with insulinoma. *Diabetologia*. 1996;39:220–225.

Scott CD, Firth SM. The role of the M6P/IGF–II receptor in cancer: Tumor suppression or garbage disposal? *Horm. Metab. Res*. 2004;36:261–271.

Seino S, Bell GI. Alternative splicing of human insulin receptor messenger RNA. *Biochem Biophys Res Commun*. 1989;159:312–316.

Sesti G, D'Alfonso R, Punti MDV, Frittitta L, Trischitta V, Liu YY, Borboni P, Longhi R, Montemurro A, Lauro R. Peptide–based radioimmunoassay for the two isoforms of the human insulin receptor. *Diabetologia*. 1995;38:445–453.

Sesti G, Federici M, Lauro D, Sbraccia P, Lauro R. Molecular mechanism of insulin resistance in type 2 diabetes mellitus: Role of the insulin receptor variant forms. *Diabetes Metab Res Rev*. 2001;17:363–373.

Sesti G, Marini MA, Tullio AN, Montemurro A, Borboni P, Fusco A, Accili D, Renato Lauro R. Altered expression of the two naturally occurring human insulin receptor variants in isolated adipocytes of non–insulin–dependent diabetes mellitus patients. *Biochem Biophys Res Commun.* 1991;181:1419–1424.

Sever R, Brugge JS. Signal transduction in cancer. *Cold Spring Harb Perspect Med.* 2015;5:pii:a006098.

Seyfried TN, Huysentruyt LC. On the origin of cancer metastasis. *Crit Rev Oncog.* 2013;18:43–73.

Sferruzzi–Perri AN, Sandovici I, Constancia M, Fowden AL. Placental phenotype and the insulin–like growth factors: resource allocation to fetal growth. *J Physiol.* 2017;595:5057–5093.

Shiratsuchi I, Akagi Y, Kawahara A, Kinugasa T, Romeo K, Yoshida T, Ryu Y, Gotanda Y, Kage M, Shirouzu K. Expression of IGF–1 and IGF–1R and their relation to clinicopathological factors in colorectal cancer. *Anticancer Res.* 2011;31:2541–2545.

Sibley CP, Turner MA, Cetin I, Ayuk P, Boyd CA, D'Souza SW, Glazier JD, Greenwood SL, Jansson T, Powell T. Placental phenotypes of intrauterine growth. *Pediatr Res.* 2005;58:827–832.

Sjögren K, Wallenius K, Liu J, Bohlooly M, Pacini G, Svensson L, Törnell J, Isaksson OG, Ahrén B, Jansson JO, et al. Liver–derived IGF–I is of importance for normal carbohydrate and lipid metabolism. *Diabetes.* 2001;50:1539–1545.

Sklar MM, Thomas CL, Municchi G, Roberts CT Jr, LeRoith D, Kiess W, Nissley P. Developmental expression of rat insulin–like growth factor II/mannose 6–phosphate receptor messenger ribonucleic acid. *Endocrinology.* 1992;130:3484–3491.

Skyler JS, Bakris GL, Bonifacio E, Darsow T, Eckel RH, Groop L, Groo PH, Handelsman Y, Insel RA, Mathieu C, et al. Differentiation of diabetes by pathophysiology, natural history, and prognosis. *Diabetes.* 2017;66:241–255.

So AI, Levitt RJ, Eigl B, Fazli L, Muramaki M, Leung S, Cheang MC, Nielsen TO, Gleave M, Pollak M. Insulin–like growth factor binding protein–2 is a novel therapeutic target associated with breast cancer. *Clin Cancer Res.* 2008;14:6944–6954.

Solomon AL, Siddals KW, Baker PN, Gibson JM, Aplin JD, Westwood M. Placental alkaline phosphatase de–phosphorylates insulin–like growth factor (IGF)–binding protein–1. *Placenta*. 2014;35:520–522.

Soussi T, Wiman KG2. TP53: An oncogene in disguise. *Cell Death Differ*. 2015;22:1239–49.

Souza RF, Wang S, Thakar M, Smolinski KN, Yin J, Zou TT, Kong D, Abraham JM, Toretsky JA, Meltzer SJ. Expression of the wild–type insulin–like growth factor II receptor gene suppresses growth and causes death in colorectal carcinoma cells. *Oncogene*. 1999;18:4063–4068.

Spampinato D, Pandini G, Iuppa A, Trischitta V, Vigneri R, Frittitta L. Insulin/insulin–like growth factor I hybrid receptors overexpression is not an early defect in insulin–resistant subjects. *J Clin Endocrinol Metab*. 2000;85:4219–4223.

Sparrow LG, Lawrence MC, Gorman JJ, Strike PM, Robinson CP, McKern NM, Ward CW. N–linked glycans of the human insulin receptor and their distribution over the crystal structure. *Proteins*. 2008;71:426–439.

Succurro E, Andreozzi F, Marini MA, Lauro R, Hribal ML, Perticone F, Sesti G. Low plasma insulin–like growth factor–1 levels are associated with reduced insulin sensitivity and increased insulin secretion in nondiabetic subjects. *Nutr Metab Cardiovasc Dis*. 2009;19:713–719.

Sudarsanam S, Johnson DE. Functional consequences of mTOR inhibition. *Curr Opin Drug Discov Devel*. 2010;13:31–40.

Šunderić M, Đukanović B, Malenković V, Nedić O. Molecular forms of the insulin–like growth factor–binding protein–2 in patients with colorectal cancer. *Exp Mol Pathol*. 2014;96:48–53.

Šunderić M, Križakova M, Malenković V, Ćujić D, Katrlik J, Nedić O. Changes due to ageing in the glycan structure of alpha–2–macroglobulin and its reactivity with ligands. *Prot J*. 2019;38:23–29.

Tan EK, Tan EL. Alterations in physiology and anatomy during pregnancy. *Best Pract Res Clin Obstet Gynaecol*. 2013;27:791–802.

Taylor SI, Cama A, Accili D, Barbetti F, Quon MJ, de la Luz Sierra M, Suzuki Y, Koller E, Levy–Toledano R, Wertheimer E, et al. Mutations in the insulin receptor gene. *Endocr Rev*. 1992;13:566–595.

Teppala S, Shankar A. Association between serum IGF–1 and diabetes among US adults. *Diabetes Care*. 2010;33:2257–2259.

Tepper OM, Capla JM, Galiano RD, Ceradini DJ, Callaghan MJ, Kleinman ME, Gurtner GC. Adult vasculogenesis occurs through in situ recruitment, proliferation, and tubulization of circulating bone marrow–derived cells. *Blood*. 2005;105:1068–1077.

Tripodi A, Branchi A, Chantarangkul V, Clerici M, Merati G, Artoni A, Mannucci PM. Hypercoagulability in patients with type 2 diabetes mellitus detected by a thrombin generation assay. *J Thromb Thrombolysis*. 2011;31:165–172.

Uhles S, Moede T, Leibiger B, Berggren PO, Leibiger IB. Isoform–specific insulin receptor signaling involves different plasma membrane domains. *J Cell Biol*. 2003;163:1327–1237.

Ullrich A, Gray A, Tam AW, Yang–Feng T, Tsubokawa M, Collins C, Henzel W, Le Bon T, Kathuria S, Chen E, et al. Insulin–like growth factor 1 receptor primary structure: comparison with insulin receptor suggests structural determinants that define functional specificity. *EMBO J*. 1986;5:2503–2512.

Uzoh CC, Holly JM, Biernacka KM, Persad RA, Bahl A, Gillatt D, Perks CM. Insulin–like growth factor–binding protein–2 promotes prostate cancer cell growth via IGF–dependent or –independent mechanisms and reduces the efficacy of docetaxel. *Br J Cancer*. 2011;104:1587–1593.

Valensise H, Liu YY, Federici M, Lauro D, Dell'anna D, Romanini C, Sesti G. Increased expression of low–affinity insulin receptor isoform and insulin/insulin–like growth factor–I hybrid receptors in term placenta from insulin–resistant women with gestational hypertension. *Diabetologia*. 1996;39:952–960.

Vander Heiden MG, Cantley LC, Thompson CB. Understanding the Warburg effect: The metabolic requirements of cell proliferation. *Science*. 2009;324:1029–1033.

Vazzana N, Ranalli P, Cuccurullo C, Davì G. Diabetes mellitus and thrombosis. *Thromb Res*. 2012;129:371–377.

Vella V, Malaguarnera R. The emerging role of insulin receptor isoforms in thyroid cancer: clinical implications and new perspectives. *Int J Mol Sci*. 2018;19:3814.

Vesel M, Rapp J, Feller D, Kiss E, Jaromi L, Meggyes M, Miskei G, Duga B, Smuk G, Laszlo T, et al. ABCB1 and ABCG2 drug transporters are differentially expressed in non–small cell lung cancers (NSCLC) and expression is modified by cisplatin treatment via altered Wnt signaling. *Respir Res*. 2017;18:52.

Wakeling EL, Brioude F, Lokulo–Sodipe O, O'Connell SM, Salem J, Bliek J, Canton AP, Chrzanowska KH, Davies JH, Dias RP, et al. Diagnosis and management of Silver–Russell syndrome: First international consensus statement. *Nat Rev Endocrinol*. 2017;13:105–24.

Walenkamp MJ, van der Kamp HJ, Pereira AM, Kant SG, van Duyvenvoorde HA, Kruithof MF, Breuning MH, Romijn JA, Karperien M, Wit JM. A variable degree of intrauterine and postnatal growth retardation in a family with a missense mutation in the insulin–like growth factor I receptor. *J Clin Endocrinol Metab*. 2006;91:3062–3070.

Wallborn T, Wüller S, Klammt J, Kruis T, Kratzsch J, Schmidt G, Schlicke M, Müller E, van de Leur HS, Kiess W, et al. A heterozygous mutation of the insulin–like growth factor–I receptor causes retention of the nascent protein in the endoplasmic reticulum and results in intrauterine and postnatal growth retardation. *J Clin Endocrinol Metab*. 2010;95:2316–2324.

Walsh CT, Garneau–Tsodikova S, Gatto GJ Jr. Protein posttranslational modifications: The chemistry of proteome diversifications. *Angew Chem Int Ed Engl*. 2005;44:7342–7372.

Walsh JH, Karnes WE, Cuttitta F, Walker A. Autocrine growth factors and solid tumor malignancy. *West J Med*. 1991;155:152–163.

Wang A, Rana S, Karumanchi SA. Preeclampsia: The role of angiogenic factors in its pathogenesis. *Physiology (Bethesda)*. 2009;24:147–158.

Wang T, Ge G, Ding Y, Zhou X, Huang Z, Zhu W, Shu Y, Liu P. MiR–503 regulates cisplatin resistance of human gastric cancer cell lines by targeting IGF1R and BCL2. *Chin Med J (Engl)*. 2014;127:2357–2362.

Warburg O. On respiratory impairment in cancer cells. *Science*. 1956;124:269–270.

Ward GM, Walters JM, Barton J, Alford FP, Boston RC. Physiologic modeling of the intravenous glucose tolerance test in type 2 diabetes: A new approach to the insulin compartment. *Metabolism*. 2001;50:512–519.

Werner H, Le Roith D. New concepts in regulation and function of the insulin–like growth factors: Implications for understanding normal growth and neoplasia. *Cell Mol Life Sci*. 2000;57:932–942.

West RC, Bouma GJ, Winger QA. Shifting perspectives from "oncogenic" to oncofetal proteins; how these factors drive placental development. *Reprod Biol Endocrinol*. 2018;16:101.

Westwood M, Gibson JM, Davies AJ, Young RJ, White A. The phosphorylation pattern of insulin–like growth factor binding protein–1 in normal plasma is different from that in amniotic fluid and changes during pregnancy. *J Clin Endocrinol Metab*. 1994;79:1735–1741.

WHO (World Health Organization), *Health topics, Cancer* [updated 2019]. Available from: (www.who.int/cancer/en/).

World Cancer Research Fund/American Institute for Cancer Research, *Colorectal cancer statistics*, [updated 2019]. Available from: (https://www.wcrf.org/dietandcancer/cancer–trends/colorectal–cancer–statistics).

Youssef A, Han VK. Low oxygen tension modulates the insulin–like growth factor–1 or –2 signaling via both insulin–like growth factor–1 receptor and insulin receptor to maintain stem cell identity in placental mesenchymal stem cells. *Endocrinology*. 2016;157:1163–1174.

Youssef A, Iosef C, Han VK. Low–oxygen tension and IGF–I promote proliferation and multipotency of placental mesenchymal stem cells (PMSCs) from different gestations via distinct signaling pathways. *Endocrinology*. 2014;155:1386–1397.

Yu H, Rohan T. Role of the insulin–like growth factor family in cancer development and progression. *J Natl Cancer Inst*. 2000;92:1472–1489.

Yu M, Yuan C, Wang H, Liu J, Qin H, Liu S, Yan Q. FUT8 drives the proliferation and invasion of trophoblastic cells via IGF–1/IGF–1R signaling pathway. *Placenta*. 2019;75:45–53.

Yuan TL, Cantley LC. PI3K pathway alterations in cancer: Variations on a theme. *Oncogene*. 2008;27:5497–5510.

Yung HW, Cox M, Tissot van Patot M, Burton GJ. Evidence of endoplasmic reticulum stress and protein synthesis inhibition in the placenta of non–native women at high altitude. *FASEB J*. 2012:26:1970–1981.

Zarrilli R, Pignata S, Romano M, Gravina A, Casola S, Bruni CB, Acquaviva AM. Expression of insulin–like growth factor (IGF)–II and IGF–I receptor during proliferation and differentiation of CaCo–2 human colon carcinoma cells. *Cell Growth Differ*. 1994;5:1085–1091.

Zawacka–Pankau J, Selivanova G. Pharmacological reactivation of p53 as a strategy to treat cancer. *J Intern Med*. 2015;277:248–259.

Zelzer E, Levy Y, Kahana C, Shilo BZ, Rubinstein M, Cohen B. Insulin induces transcription of target genes through the hypoxia–inducible factor HIF–1alpha/ARNT. *EMBO J*. 1998;17:5085–5094.

Zeng L, Perks CM, Holly JM. IGFBP–2/PTEN: A critical interaction for tumours and for general physiology? *Growth Horm IGF Res*. 2015;25:103–107.

Zhang YK, Wang YJ, Gupta P, Chen ZS. Multidrug resistance proteins (MRPs) and cancer therapy. *AAPS J*. 2015;17:802–812.

Zhou P, Ten S, Sinha S, Ramchandani N, Vogiatzi M, Maclaren N. Insulin receptor autoimmunity and insulin resistance. *J Pediatr Endocrinol Metab*. 2008;21:369–375.

Zhou R, Diehl D, Hoeflich A, Lahm H, Wolf E. IGF–binding protein–4: Biochemical characteristics and functional consequences. *J Endocrinol*. 2003;178:177–193.

In: A Closer Look at Membrane Proteins
Editor: Tristan B. Møller

ISBN: 978-1-53618-149-4
© 2020 Nova Science Publishers, Inc.

Chapter 2

SIMULATING MEMBRANE PROTEINS

N. K. Roy[1,2,*]

[1]ITS-Research Computing, Northeastern University,
Boston, MA, US
[2]The Charles Stark Draper Laboratory, Inc. (Draper),
Cambridge, MA, US

ABSTRACT

Membrane proteins may contain a significant portion of their mass within the interior of the membrane or are only associated to the membrane surface. The transmembrane (TM) part can be helical or have a sheet topology. The TM part can have a variety of sizes, molecular weights and conformations. Membrane proteins govern biological processes such as energy conversion, transport, signal recognition and transduction. Up to 30% of the encoded proteins in the genome of all organisms are such proteins, and are 60% of all drug targets. Currently less than 1% of the protein structures deposited in the RCSB Protein Data Bank are membrane proteins. Simulating membrane proteins correctly provides the best way to study them. Ion channels are a class of membrane proteins where the passive transport is influenced by the membrane potentials. Many such ion channels have the selectivity filter

---

[*] Corresponding Author's E-mail: nroy@draper.com.

and gating mechanism embedded in the membrane core. Asymmetric ion concentrations across the membrane also affect transport and protein functions. These are difficult to study. Large scale molecular dynamics (MD) with coarse graining in both the membrane lipid bilayer and in parts of membrane protein itself is generally the method used in any simulation study to understand the mechanisms of such proteins' structure-function relationships and dynamic modes. Principal component analysis of the protein is also frequently used. This chapter gives an overview of the current methods used to prepare and study membrane proteins. The focus is on large scale simulations with special emphasis on scalable parallel methods. Correctly relating molecular structures to the physiological properties of the protein is a major challenge in the field. All the effects of the inhomogeneous lipid bilayer, potentials, ion/anion concentrations, that cover both spatial and temporal scales must be included. This has challenges when systems have thousands of explicit atoms and require simulations on the micro-second scale. We review these challenges and explain methods that have been used to overcome the short comings of explicit MD simulations.

Keywords: membrane proteins, transmembrane, genome, molecular dynamics, ion/cation concentrations, membrane potentials, force fields, semi-explicit methods, lipid, membrane curvature

INTRODUCTION

Molecular dynamics (MD) simulation is a molecular mechanics method that is implemented using numerical methods studying the motion of a system of particles (atoms, molecules, entities) under the influence of internal and external forces [1]. These forces are interactions between the particles and due to other parameters like temperature, pressure, and additional constraints [2]. Additional constraints include forces in steered or targeted MD. The empirical potential energy function that relates structure to energy and describes the forces between atoms using harmonic and periodic potentials to model covalent bond mediated interactions, as well as Coulomb and Lennard Jones-like potentials to represent electrostatic and van der Waals interactions, form the basis of MD force calcuations [3]. These forces are called "force fields". Typically calculated in a few femtosecond time steps they predict how each atom will move. Repeating the time step millions of times, a trajectory of all atoms in the system over time is generated that permits studying the dynamics of the (membrane) protein of interest and its microenvironment at a level of detail not accessible by experiments. [4-5]. By parallelizing the system using a large compute cluster of computers it is now possible to simulate a range of system sizes upto several million atoms even on a millisecond time scale [6-8]. Several force fields have been developed like AMBER, CHARMM, and GROMOS [9-11]. Typically in MD simulations of very large explicit systems, it is well known that there are several problems. These include exponentially increasing equilibrium and non-equilibrium relaxation times, correlation times and lengths [12]. In very large systems there are problems like critical slowing down and other finite size effects that can take on special significance [13]. Further with any simulation the full characterization must have reliable error estimates. This can only be obtained after several runs from a number of different initial starting configurations. Figure 1 gives an example of time and length scales of different computational methods. In any MD system considerations must also include the ensemble to be used – for example NVE (micro-canonical), NPT or NVT [14]. An important consideration here is if diffusion is needed. Then the density varies, and a grand-canonical enbsemble may be more favorable over the canonical NPT [15]. NPT is generally also preferable if you want to equilibrate the system. NVT is then used to collect the statistics of interest.

Another important consideration is creating the starting structures and system of MD simulations. The membrane protein environment is characterized by two different chemical regions that have to be modelled correctly by the force field [16]. This is done using implicit or explicit representations of the lipid and water phases. In general if the reported protein structure is used there are many defects. These include overlapping atoms, missing hydrogens, improper N and P terminus and so on. If modeled the total energy of the system would explode. The entire protein with the lipid membrane, water, and ion/counterions has to be systematically relaxed, overlapped and equilibrated. For water soluble proteins, generation of the initial system is completed by solvating the system with a water/ion solution [17]. However, membrane protein simulations require the additional working step of accommodating the protein in the bilayer. Here either the bilayer is constructed around the protein, or the protein is inserted into a pre-equilibrated bilayer. As shown in Figure 2 there are a number of tranmembrane protein toplogies. Depending on the protein's shape, its cavity structure in the transmembrane section, the number of membranes spanned by the protein and the membrane curvature (plane bilayer versus vesicles or micelles), individual challenges arise that have to be taken into account by both accommodation strategies [18]. Table 1 gives the times and sizes of some MD simulations of membranes [19-22]. These larger scale models also enable studies of the collective behaviour of multiple copies of membrane proteins, such as the influence of crowding of membrane proteins on their clustering and diffusion.[23-24]. The dynamic properties of membranes play a key role as regulatory mechanisms and will influence the mechanical properties of cell membranes.

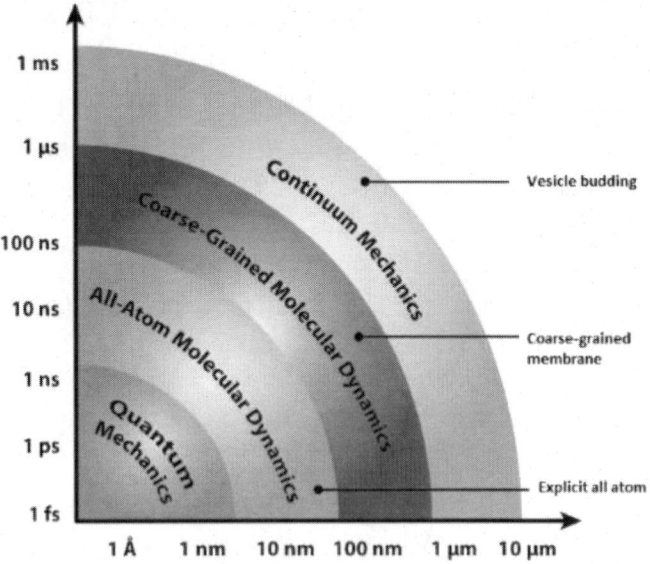

Figure 1. Computational methods over a range of length and time scales in membrane protein simulations.

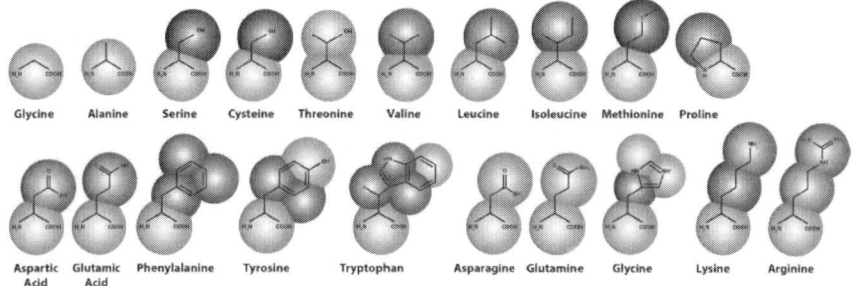

Figure 2. Martini coarse-grained model extension to amino acids, colored by bead type (where purple is apolar, blue and green are intermediate, gray and orange are polar and red represents charged particles).

Table 1. For each MD system, granularity of the simulation (atomistic versus coarse-grained), the number of atoms/particles (including water) in the simulation system, the duration of the production run simulation, the approximate linear dimension of the simulation box, and the resultant trajectory file size are given. The last column gives the Reference

MD System	Type	No of Entities	Time Scale	Length Scale	Size	Ref
Aqp0 in a PC bilayer	Atomistic	156K atoms	100ns	10nm	5GB	[19]
EphA2 in a PC/PG bilayer	Coarse grained	292K particles	10μs	30nm	100GB	[20]
144 GPCRs in a PM model	Coarse grained	2.6M particles	10μs	125nm	1.1 TB	[21]
Influenza A virion10	Coarse grained	5M particles	5μs	90nm	1.9 TB	[22]

COARSE GRAINED GRAINED MD SIMULATIONS

In coarse grained models the length-scale at which chemical components are modeled become important. Such a model necessarily lumps many atomic degrees of freedom into a single coarse-grained bead. As with the MD approach the coarse grained molecular model (CGMD) treats molecules classically. Newton's law of motion are integrated according to potentials, which define the forces beween each bead in the system.

$$m_i \frac{\partial^2 r_i}{dt^2} = F_i, F_i = -\frac{\partial V}{\partial r_i}, i = 1 \ldots N \tag{1}$$

Eq 1 describes the motion of N particles, each with mass, m_i, experiencing a force, F_i, due to a potential energy function, V, itself a function of the configuration of all atoms in the system that are close enough to exert a measurable force. NAMD [25] integrators is one software package that can be used for CGMD to perform simulations. MD simulations make contact with observables, like temperature and pressure. Temperature is defined by the kinetic energy of the particles, while macroscopic pressure is defined by the average of the molecular virial given in Eq 2, and Eq 3 respectively.

$$\tfrac{1}{2} N_{df} k_B T = E_{kin}, E_{kin} = \tfrac{1}{2} \sum_i^N m_i \mathbf{v}_i \cdot \mathbf{v}_i \tag{2}$$

$$\mathbf{P} = \tfrac{2}{V}[E_{kin} - \Lambda], \Lambda = -\tfrac{1}{2}\sum_{i<j} \mathbf{r}_{ij} \cdot \mathbf{F}_{ij} \tag{3}$$

V is the volume of the system, E_{kin} is the kinetic energy, $\mathbf{r}_{ij}$ is the distance vector between particles, i and j, $\mathbf{F}_{ij}$ is the corresponding force, N_{df} is the number of degrees of freedom (3N - 3 for N particles, minus any constraints) and Λ is the virial. The choice of these forces and the physical quantities they represent-dispersion forces, electrostatics and bonded forces-define themodel and determine its ability to reproduce observed physical phenomena. CGMD can be done on models built using structure-based, force-based and energy-based force-fields. Because coarse-graining requires a simplification of many degrees of freedom, it is impossible to build a model that simultaneously reproduces the all of the geometric, thermodynamic and kinetics features of a physical system. To build a coarse-grained model, it is therefore necessary to choose which physical properties are essential to the behavior of the target system. Examples of such systems are reviewed extensively [26-29].

In the "Center for Molecular Modeling Coarse-Grained" (CMM-CG) [26] model the entire structural properties of dimyristoylphosphatidylcholine (DMPC) bilayer is reproduced. The CMM-CG model maps three water molecules onto a single bead. Non-bonded forces are modeled with general Lennard-Jones (LJ) potentials with a potential well depth and zero-position, which is tuned to reproduce the desired structure and thermodynamic properties of the target system. The softer 12-4 potential is used to model dispersion forces in water by matching the melting temperature, density and vapor pressure observed in bulk and thin-film test simulations. A potential of mean force (PMF) between CG beads is then estimated using pair correlation functions, or a radial distribution functions (RDF).

In the case of the "Force Matching with the Multiscale Coarse Grained" (MS-CG) [27] model, force-matching to develop a rigorous coarse-grained force field directly from forces measured in all-atom simulations was used. This is a variational method in which a coarse-grained force field is systematically developed from all-atom simulations under the correct thermodynamic ensemble. It is now possible to develop the exact many-body coarse-grained PMF from a trajectory of atomistic forces with a sufficiently detailed basis function.

Introducing protein detail to a coarse-grained force field requires an accurate model for both the structure and dynamics of the protein itself, as well as the interactions with surrounding lipids and solvent. One example of such coarse graining is in the so called "Martini Proteins" [30]. In the Martini force field, amino acids are mapped onto as many as five beads (Figure 2), one of which represents the polypeptide backbone. Residues with rings (His, Phe, Tyr, Trp) use a finer mapping and improper dihedral terms to preserve the topology of these rings. Intra-amino acid bonded potentials-bonds, angles and dihedrals-have equilibrium values equal to the average of distributions measured from all bonded amino acid pairs found. These are sorted by helix, coil and extended secondary structure, as measured by the DSSP ("define secondary structure of proteins") prediction algorithm [31], so that the Martini model includes the effect of secondary structure on the apparent hydrophobicity and polarity of its constituent particles. This secondary structure remains fixed through the simulation. Thus the Martini model cannot sample secondary structure changes. It is possible to reconstitute atomistic details from a coarse-grained simulation using a "back-mapping" procedure similar to simulated annealing [32]. By tuning these models it is possible for CGMD to accurately explain protein-bilayer interactions, peptide self-assembly and protein binding. These methods can also model internal structural changes that guide the biological functions of many proteins.

ASYMMETRIC ION CONCENTRATIONS

Differences in ion concentrations across the membrane that are established under the action of various membrane transport proteins can give rise to a difference in electric potential. Reproducing this set of conditions in computer simulations is nontrivial. Several methods have been used succesively to simulate this effect. To allow for the simulation of ion channels with a realistic implementation of asymmetric ion concentration and transmembrane potential boundary conditions, a grand canonical Monte Carlo (GCMC)/Brownian dynamics (BD) was implemented in one case [33]. Here asymmetric boundary conditions were imposed on a finite nonperiodic simulated system surrounded by concentration buffer regions. Insight into the factors governing the permeation of wide aqueous pores was possible. Imposing asymmetric concentrations in explicit solvent MD is hard. Such explicit solvent MD simulations are normally performed with conventional periodic boundary conditions (PBCs), which are critical to reduce finite-size effects. Unavoidably, the PBCs also eliminate the distinction between the two sides of a membrane. Because there is a single continuous bulk solution where ions are free to diffuse and equilibrate, concentration gradients across the membrane cannot be simulated. To overcome this, simulation of asymmetric ion concentrations in MD simulations with explicit solvent using a dual-membranes–dual-volumes strategy was tried [34]. Two spatially separated membranes are included to create two disconnected bulk phases between them. In a more recent effort one of the two membranes is replaced by an artificial vacuum separator to reduce the computational burden [35]. In another case manually adjusting the number of cations and anions in the two bulk regions makes it possible to set the effective membrane potential near some pre-chosen value V_m [36]. These methods serve to increase computational costs. Other attempts include introducing energy steps at the boundaries of the periodic cells that separates the two solutions and generates a nonuniform distribution of the solute molecules across the cell boundaries with assymetric external fields. These result in a net charge imbalance across the membrane [37]. There is no simple relationship between the energy step, the charge imbalance, and the resulting membrane potential.

INTERFACE OF MEMBRANES

One interesting set of problems concerns the mechanism by which small peptides and peripherally associated membrane proteins bind to and interact with the water/membrane interface. Studies include membrane lytic toxins, model peptides, fusion proteins, and peripherally associated signal transduction proteins and biosynthetic enzymes. For example in one study [38] implicit solvent calculations was used on cobra cardiotoxin CTX A3 to interpret polarized attenuated total internal reflection infrared spectroscopy data, suggesting modes of binding of this toxin to zwitterionic and anionic membranes. This toxin appeared to have a greater thinning effect on anionic phosphatidylglycerol monolayers than on those composed of zwitterionic phosphatidylcholine species.

A number of other membrane–water interface associated peptides have been investigated. In one study with the HIV fusion protein gp41 and mutants in POPE bilayers [39]comparison between simulations and attenuated total internal reflection infrared spectroscopy was used to determine the orientation of peptides relative to the bilayer. The simulation component of this study involved the removal of several lipids from one leaflet in order to accommodate the protein inclusion.

Several larger peripheral membrane proteins have been investigated by molecular dynamics. One study looked at cytochrome c in association with different alkanethiol self-assembled monolayers, with the aim of understanding the structural features of the protein in these complexes and the nature of the monolayer association. Many peripherally associated proteins have important roles in signal transduction and disease, and simulations continue to be a powerful method to understand the detailed lipid–protein interactions responsible for their activity.

Figure 3 shows an example of the membrane anchored protein prostaglandin H2-synthase. Its substrate, arachidonic acid, is a fatty acid in the membrane and cannot be found in cytoplasm. The only way to effectively study such a protein is via large scale simulations.

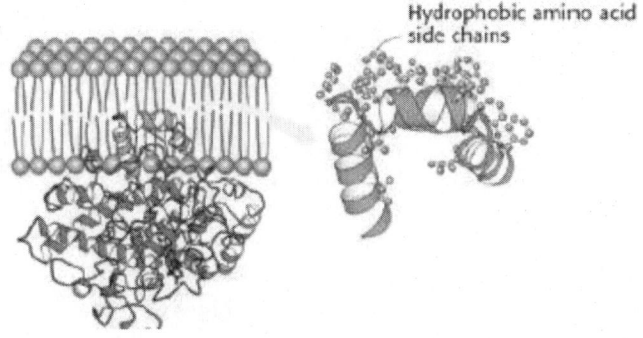

Figure 3. The membrane anchored protein prostaglandin H2-synthase.

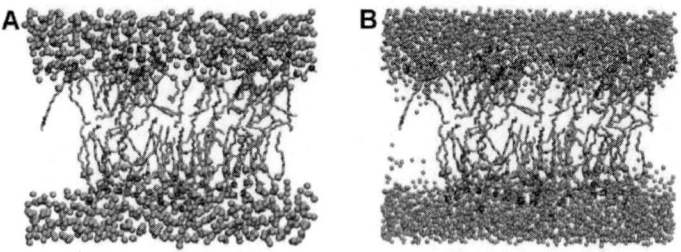

Figure 4. Hydration of the heads group region in both coarse-grained (A) and all-atom models (B) of a membrane protein. Lipids are removed.

Figure 4 shows an example of hydration of the heads group region in both coarse-grained (A) and all-atom models (B) of a membrane (lipids are removed). After large scale MD simulations reverse coarse-graining is done mapping back coarse-grained beads to the all-atom clusters, and resolvating the system. Minimization steps with simulated annealing while constraining atoms to the postion of the correspondning coarse-grained beads completes the process.

TRANSMEMBRANE SIMULATIONS

It is known that membrane spanning helices can be considered the smallest autonomous membrane protein domains. Simulations can be done in order to study properties of fundamental interest, such as the dynamics of isolated helices [40], protein–lipid interactions [41], helix association [42], and the behaviour of simple channels such as those formed by certain fungal proteins and toxins [43]. MD simulations have also looked at the effects of hydrophobic mismatch on peptide and bilayer dynamics. Mismatch results when the effective hydrophobic thickness of the bilayer does not match that of a perfectly transmembrane-oriented helix [44]. Several mechanisms might compensate for this mismatch, including changes in peptide tilt, changes in secondary structure, peptide association, aggregation of different lipid species near the protein, and adaptation of bilayer structural properties like thickness or curvature. The dynamic properties of individual helices can be examined in detail with MD. A number of proline and glycine-containing sequence motifs have been extensively studied. Transmembrane helix flexibility mediated by specific sequence motifs have important consequences for the activity of membrane proteins such as gated ion channels [45-50]. The number of transmembrane helices in membrane proteins ranges from 1 to 14. Single transmembrane helix in bitopic membrane proteins are the most numerous. The second most numerous transmembrane proteins are those with seven transmembrane helices [51]. This is followed by proteins with 2, 4 and 12 transmembrane helices.

The dynamic properties of individual helices can be examined in detail with MD. Ion-channel pore helices like proline-linked helices have been extensively studied. Proline and glycine are both thought to be important mediators of hinge-bending motions. Proline disrupts normal helical hydrogen bonding and participates in repulsive steric interactions with adjacent backbone atoms. Study on TMH-2 of the chemokine receptors identified a highly conserved TXP sequence motif by multiple sequence alignment [52]. MD simulations were employed to investigate the effects of this sequence on the behavior of polyalanine in a hydrophobic environment. This indicated that the hydroxyl-containing amino acids also modulate the kink behavior of proline-containing sequences.

The interaction between α-helices is thought to be one of the most important determinants of membrane protein structure and function [53]. Proteins comprising pairs of α-helices can be employed as models for understanding these interactions. To get the interactions between pairs of helices simulated annealing and global searching molecular dynamics [54] or Monte-Carlo simulations [55], either with an all-atom MD force field [56] or a simplified interaction potential function [57] was done.

Proton transport presents a special challenge to molecular dynamics simulations because protons move between different water molecules and are not easily treated by a classical potential function. Viral and other small ion channels from small proteins (60-120 amino acids) are many such membrane proteins that have proton transport. Gramicidin A [58], the influenza A M2 channel [59], and the engineered LS2 channel are examples. Different approaches have been developed for studying proton transport in membrane proteins. One method involves using the PM6 water model. PM6 is a polarizable and dissociable empirical water model consisting of O^{2-} and H^+ units [60]. The empirical valence bond (EVB) theory has also been used to model LS2 channel [61]. In all cases extensive MD simulations of a multitude of parameters and potential models are done to determine the best fit to experimental observations. These are all areas in which large-scale MD simulations do not provide correct answers.

Another area where explicit MD simulations need to be modified is in the proton exclusion problem, as seen in aquaporins (AQP). These membrane proteins allow water and glycerol to diffuse through while excluding protons. This would destroy the proton electrochemical gradient and starve cells to death. Several models for proton exclusion have been proposed [62-66] and used with classical and steered (targeted) MD simulations. In simulations looking at water permeation, key structural features including the NPA motif, a constriction region (also termed ar/R), and the helix dipoles have been identified as contributors to this specificity. Methods to quantify water conduction properties such as osmotic permeability of AQP via MD simulation methods have also been developed. These methods involve inducing a hydrostatic pressure difference across a membrane with embedded AQP [67].

In the case of mechanosensitive transmenbrane channels [68] a combination of MD simulations with normal mode analysis has been successively used [69-71]. However typically such simulations provides more than one mechanism for opening and closing. These multiple pathways provide a challenge in the absence of crystal structures on either the open or closed states for guidance. High-resulution structural methods cannot access these states. The crystal structures obtained of such states will most likely be a mutant one based on modeling. Figure 5 shows the two open states of the mechanosensitive ion channel MscL [72].

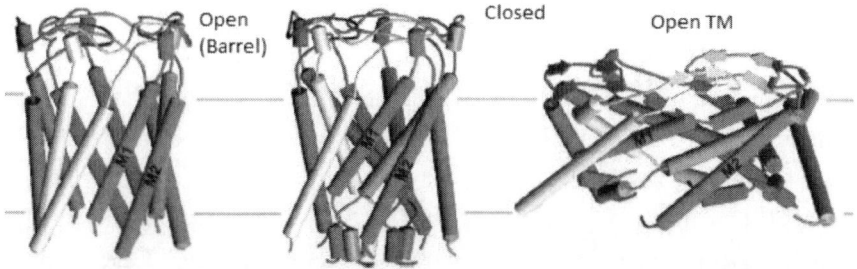

Figure 5. Two possible open states of the mechnosensitive ion channel MscL. MD simulations can predict both types of gating where the pore diameter of one open state (Barrel ~ 15 Å) is nearly half that of the other (Open TM pore diameter ~ 30Å).

Direct MD simulations however are shown to very effective in the case of ABC-type transporters [73]. As one of the largest superfamilies of proteins [74] they are involved in multidrug resistance in cancer cells and bacteria and in genetic diseases such as cystic fibrosis. They are present in eukaryotic and prokaryotic systems and have the characteristic LSGGQ signature motif of their ATPase domain. Using targetted molecular dynamics (TMD) it has also been shown how MD can be used for structural modeling. The first reported crystal structure of a "complete" ABC-transporter, MsbA from E. coli (ECMsbA), was resolved at 4.5Å resolution [75]. This structure revealed only the C-α polypeptide trace of the protein, and coordinates for a significant part of the nucleotide binding domain (NBD) consisting of 78 residues including the conserved ATP binding "Walker A" motif could not be determined due to disorder. Starting from the C-α polypeptide trace, the backbone and side chain atoms were generated, and the structure of the missing part of the NBD was modeled based on homology with known high-resolution structures of the NBDs of other ABC transporters. MD simulations were subsequently used to test the stability of the model. While the monomer was stable to MD simulations the dimer was not. Reorienting the transmembrane domains of ECMsbA with respect to the NBDs in order to transform the "back-to-back" dimer into a "head-to-tail" model a suitable template was found to model another ABC-transporter P-glycoprotein [76].

Another avenue used to simulate membrane proteins is biased molecular dynamcis. Using the biasing forces on ATP synthase creates non-equlibrium MD simulations. Steering or acceleration may also be used. This is favored over the reqular equilibirum MD. Events of interest like ATP binding or release occur on the millisecond time scale but the large size of the protein limits simulation to the nanosecond. One example is a biasing force to cause 120° rotations of the cental stalk. The drawback with this method is that key relaxation events may be missed when a process that occurs in the millisecond time scale in nature is forced to occur on the nanosecond time scale accessible to protein MD simulations. Biased molecular dynamics simulations thus do not yield a full transition path. However, useful mechanistic information can still be derived from such studies [77].

Finally processes involving the breaking of chemical bonds cannot be studied uisng MD. A quantum mechanical (QM) treatment is needed. QM is computationally expensive and currently limited to a few hundred atoms. The QM/MM (molecular mechanics) method is a technique used to study such systems. The reactive centre and its immediate surroundings are modeled in electronic detail using a quantum mechanical approach while the rest of the system is treated classically to atomic detail using molecular dynamics. NAMD 2.12 and later [78] has this feature. Hybrid QM–MM simulations in NAMD divide the system into MM and QM regions, using a classical force field to treat the classical atoms and passing the information that describes the quantum atoms in the system to a quantum chemistry package, which is expected to calculate forces for all QM atoms, as well as the total energy of the QM region and the partial charges. All bonded and nonbonded interactions among MM atoms are handled by NAMD's CHARMM force field, whereas all interactions among QM atoms are handled by the quantum chemistry package in its chosen theory level. QM/MM methods can now be used for mechanical and electrostatic embedding, treatment of covalent bonds, link atoms, and point charge alteration and redistribution.

OUTER MEMBRANE PROTEIN SIMULATIONS

β-barrel membrane proteins are unique to the outer membranes of mitochondria, chloroplasts, and gram-negative bacteria. The bacterial proteins, which range in size from 8 to 22 β-stranded barrels, perform a variety of functions, from enzymatic lipid cleavage to nutrient uptake functions to iron transport [79]. MD simulations provide a useful tool for exploring the dynamics here, and their interactions with their environment. In OmpA MD has been used to explore the behaviour of the protein in three different environments: a lipid bilayer, a detergent micelle and a (detergent-containing) crystal [80]. Figure 6 shows the structure at the beginning and end of a 25ns simulation for a similar protein OmpX inserted in lipid DMPC with water and equilibrated. In this case the simulation was stable with very small changes in the lipid layer. Water molecules trapped in the β-barrel due to hydrogen bonds are seen. There is large structural fluctutions on the extracelluar part of the protein. MD simulations effectively can study this. The problem is to sample adequately the conformational space accessible for the species of interest. In this case it may the question of why water cannot pass through. In another case it may be the translocation mechinism and kinetics of the siderophore through FhuA [81]. The sampling of the conformational space when there is a large conformational change in the protein in the lipid layer can be difficult in MD runs at the "ns" scale. While using several other methods like applying forces or pulling parts of the protein may result in inducing conformational changes in such simulations, in many instances other mechnism are at work that cannot be studied via MD in such ways. For example, a significant conformational change in the plug domain of FhuA is required for the siderophore to either passively diffuse or be translocated into the periplasm. To study this by MD simulations requires long runs (ms scale) to trace the confomational changes as the diffusion occurs.

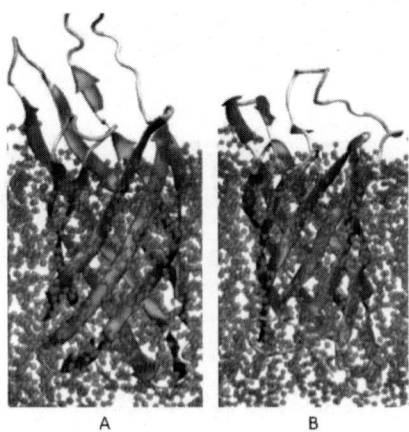

Figure 6. OMP-X in a lipid bilayer. The end of the MD simulation (B) after 25 ns starting from initial equilibriated configuration (A). Waters are removed – however water is trapped in the protein via hydrogen bonds. See text for more details.

CONFORMATIONAL CHANGES IN SIMULATIONS

In many classes of membrane proteins like the voltage gated K channel KcsA (α-helical protein) or KvaP [82] the gating mechanism results in conformational changes (range of states) in the embedded region of the protein. Particularly in the selectivity filter regions. Here the large reaction force field in the pore region due to an extensive hyper-polarized/depolarized external membrane potential needs to factored in. Figure 7 shows a system [83] where an extended dielectric region was used to get the reaction force in the pore and the corresponding conformational changes induced in KcsA in the pore. Simulations were 2ns, and the entire system had ~ 60K explicit atoms of a single membrane protein, lipids (5 shells), water and ions. Interhelix dynamics and force field maps can then help understand the gating and related conformational changes.

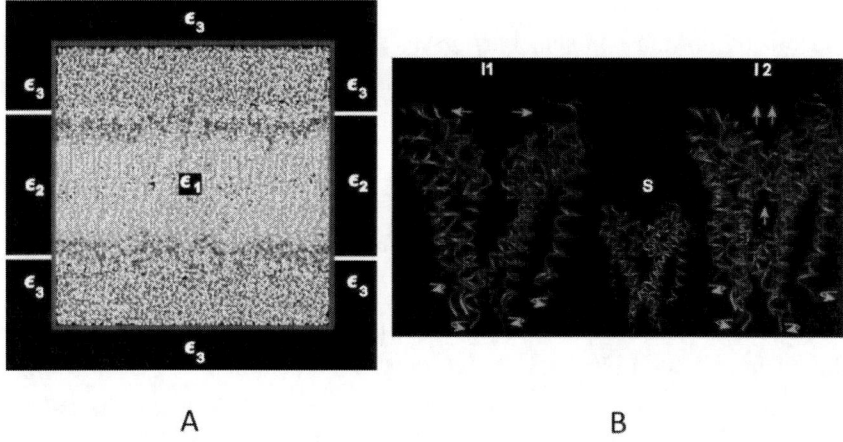

Figure 7. KcsA in DPPC surrounded by a dielectric continuum. The red region is the Helmholtz layer (A). In B with water and lipids removed the effects of the reaction force field in seen in conformational changes in KcsA. Straight green arrows show the relative translational motions between the chains from C (closed) to O (open) that occur with the rotations of each of the four chains in KcsA shown with the curved arrows. These changes are reversible See text for details.

Some membrane proteins like the bacterial leucine transporter (LeuT), a homologue of the eukaryotic Na^+/Cl^--dependent neurotransporters responsible for terminating synaptic transmission by driving the cellular uptake of neurotransmitters, including the biogenic amines, have a rare conformational change event [84]. A large number of molecular dynamic steps are necessary for this event to occur, which allow a system to overcome energy barriers and conformationally transition from one potential energy minimum to another. Using a combination of accelerated MD (aMD) and principal component analysis (PCA), protein segments that are most involved in structural changes can be identified. The RMSF (root mean square fluctuations) calculation can be used to determine how much each residue moves during the trajectory. aMD simulations are routinely performed to assess time-dependent protein conformational change [85] and are fully integrated into commonly used software packages including NAMD [86] and Amber [87]. These results can be then compared to experimental structures.

Cell membranes present barriers to the permeation/diffusion of polar molecules. Membrane transport proteins have evolved to facilitate the passage of specific molecules across this barrier. An example is lactose permease (LacY) [88]. LacY performs the symport of lactose or other galactoside molecules with H^+ in a 1:1 stoichiometry. MD simulations of LacY in the apo state (without bound TDG) were run to discover whether substantial conformational changes would be observed (on a 25ns timescale). Further from such runs the goal was to explore the possible relationship of such changes to the transport mechanism of LacY (longer timescale of 100 ms). Several references provide excellent results [89-91] using a variety of techniques, but based on the origianal classical MD approach.

CONCLUSION

MD simulations have contributed significantly to the understanding of protein structure-function relationships. MD has enabled an atomic-level description the interactions of channel-forming proteins with the solutes whose transport they mediate, characterization of ligand-binding processes, and in case of voltage gated membrane proteins insights into the selectivity filter and gating mechanisms. Simulations of outer membrane proteins have provided information on the protein dynamcis, role of the extracelluar structures and the relationship with the lipid layer embedded proteins in conferring selectivity to porins, or in water pathways. Mechanosensitive channel MD simulations have been used to study the relative flexibility throughout the protein, as a means by which insights into the mechanism of gating can be gained. MD simulations now contribute to the understanding of the structural, dynamical, and functional properties of receptors. Extensive work is done to compare the structureal dynamics with native crystal forms eluciadiated from ligand-free and bound forms. These include crystal structures from X-ray, neutron, NMR, EPR studies and those augmented by threading and holomology modeling [92]. The MD approach also allows for multiple random conformations to be taken at a time. These are validated after long equilibraiting runs using experimental data. Here the experimental data is serving as limitations to be placed on the conformations (e.g., known distances between atoms). Only conformations that manage to remain within the limits set by the experimental data are accepted. This approach often applies large amounts of experimental data to the conformations which is a very computationally demanding task, but will result in valid conformations of open and closed states of membrane proteins that cannot be isolated by crystallization.

Methodological limitaitons do exist in large scale atomistic simulations of membrane proteins. Much of this relates to the force field used and the thermodynamic ensemble. While classical force fields (GROMOS, CHARMM, AMBER) have limited implementations of electronic polarizablity [93] others like reactive force fields and coarse-grained force fields have been shown to work well in many situations [94]. Similarly there are a number of water models that can be used with ions/counter ions [95]. There is also the question of the optimal method to embed a complex membrane protein within a lipid bilayer. In fact, the effect of the lipid composition of the bilayer model on the dynamics of membrane proteins is well established [96] and needs to considered carefully before undertaking large-scale MD runs. All these limitations can however be overcome in judicious choices of the methods to use before undertaking large-scale MD runs.

Parallel techniques used in MD codes have now made possible the use of MD simulations for the routine study of systems consisting of thousands of atoms for multi-nanoseconds. On the temporal and spatial scales there are a number of events that can occur. These can now be studied in more detail. The typical timescale for allosteric effects is usually from microseconds to milliseconds. During such complex events, the transitions between two stable states are separated by high free-energy barriers. Modified MD techniques are used for the study of such events. Coarse-grained models, steered MD, biased MD approaches, targeted MD, and alchemical free-energy pertubations. Another effect that is now being studied is the curvature of the lipid bilayer by using large systems that can be parallelized and a elastic curvature force field. In addition a flexible surface model (FSM) can be used. These enhancements are added to the usual large-scale MD simulations to include curvature stress.

The growth in the number of high-resolution membrane protein structures is encouraging. Simulations can provide additional dynamic details of the static snapshots given by membrane protein structures and can be used to investigate and build models of other conformational states. Models and simulations of open and closed states of the membrane proteins, conformational substates of transporters, and dynamics linking the crystal structures of the different states will help in designing drugs, inhibitors, and understanding cell signaling and ligand docking. The ultimate goal of memnbrane MD simulations is to be able to do realistic simulations. Here effects of a controlled pH, metabolites and ions/counterions with a constant energy flow, and a wide variety of embedded and peripheral proteins are included with a actin skeleton. These billion atom explicit systems now model the plasma membrane in full complexity.

REFERENCES

[1] Chandler, D. (1987). *Introduction to Modern Statistical Mechanics.* Oxford University Press, New York, NY.

[2] Lindahl, E., Hess, B. & van der Spoel, D. (2001). GROMACS 3.0: A package for molecular simulation and trajectory analysis. *J. Mol. Model.*, *7*, 306-317.

[3] MacKerell, A. D. Jr. (1998). All-atom empirical potential for molecular modeling and dynamics Studies of proteins. *J. Phys. Chem. B*, *102*, 3586-3616.

[4] Ingo´lfsson, H. I., Arnarez, C., Periole, X. & Marrink, S. J. (2016). Computational "microscopy" of cellular membranes. *J Cell Sci.*, 1-12.

[5] Stansfeld, P. J. & Sansom, M. S. P. (2011). Molecular simulation approaches to membrane proteins. *Structure*, *19*, 1562-1572.

[6] Klepeis, J. L., Lindorff-Larsen, K., Dror, R. O. & Shaw, D. E. (2009). Long-timescale molecular dynamics simulations of protein structure and function. *Curr. Opin. Struct. Biol.*, *19*, 120-127.

[7] Allen, F., Almási, G., Andreoni, W., Beece, D., Berne, B. J., Bright, A., Brunheroto, J. & Cascaval, C. (2001). Blue gene: A vision for protein science using a petaflop supercomputer. *IBM Syst. J.*, *40*, 310-327.

[8] Rapaport, D. C. (2004). *The Art of Molecular Dynamics Simulation.* Cambridge University Press, Cambridge, UK.

[9] Case, D. A., Cheatham, T. E., Darden, T., Gohlke, H., Luo, R., Merz, K. M., Onufriev, A., Simmerling, C., Wang, B. & Woods, R. J. (2005). The Amber biomolecular simulation programs. *J. Comput. Chem.*, *26*, 1668-1688.

[10] Patel, S., MacKerell, A. D. Jr. & Brooks, C. L. III. (2004). CHARMM fluctuating charge force field for proteins: II-Protein/solvent properties from molecular dynamics simulations using a nonadditive electrostatic model. *J. Comput. Chem.*, *25*, 1504-1514.

[11] Christen, T., H"unenberger, P. H., Bakowies, D., Baron, R., Burgi, R., Geerke, D. P., Heinz, T. N., Kastenholz, M. A., Krautler, V., Oostenbrink, C., Peter, C., Trzesniak, D. & VanGunsteren, W. F. (2005). The GROMOS software for biomolecular simulation: GROMOS05. *J. Comput. Chem.*, *26*, 1719-1751.

[12] Stanley, H. E. (1971). *Introduction of Phase Transitions and Critical Phenomena*, Oxford University Press, New York.

[13] Ceriotti, M., Bussi, G. & Parrinello, m. (2010). Colored-Noise Thermostats à la Carte. *J. Chem. Theory Comput.*, *6*, 1170-1180.

[14] Leach, A. R. (1996). *Molecular Modelling. Principles and Applications*. Addison Wesley Longman, Essex, England.

[15] Frenkel, D. & Smit, B. (1996). *Understanding Molecular Simulations. From Algorithms to Applications*. Academic Press, San Diego, California.

[16] Allen, M. P. & Tildesley, D. J. (1987). *Computer Simulation of Liquids*. Oxford University Press, New York.

[17] Marrink, S., Devries, A. & Tieleman, D. (2009). Lipids on the move: Simulations of membrane pores, domains, stalks and curves. *BBA-Biomembranes*, *1788*, 1, 149-168.

[18] Van der Ploeg, P. & Berendsen, H. J. C. (1983). Molecular-Dynamics of a Bilayer-Membrane. *Mol. Phys.*, *49* (1), 233-248.

[19] Stansfeld, P. J., Jefferys, E. & Sansom, M. S. P. (2013). Multiscale simulations reveal conserved patterns of lipid interactions with aquaporins. *Structure*, *21*, 810-819.

[20] Chavent, M, Seiradake, E, Jones, E. Y. & Sansom, M. S. P. (2015). Structures of the EphA2 receptor at the membrane: role of lipid interactions. *Structure*, *24*, 337-347.

[21] Koldsø, H. & Sansom, M. S. P. (2015). Organization and dynamics of receptor proteins in a plasma membrane. *J. Am. Chem. Soc.*, *137*, 14694-14704.

[22] Reddy, T., Shorthouse, D., Parton, D. L., Jefferys, E., Fowler, P. W., Chavent, M., Baaden, M. & Sansom, M. S. P. (2015). Nothing to sneeze at: a dynamic and integrative computational model of an influenza A virion. *Structure*, *23*, 584-597.

[23] Janosi, L., Li, Z., Hancock, J. F. & Gorfe, A. A. (2012). Organization, dynamics, and segregation of Ras nanoclusters in membrane domains. *Proc. Natl. Acad. Sci. USA*, *109*, 8097-8102.

[24] Chavent. M., Reddy, T., Goose, J., Dahl, A. C. E., Stone, J. E., Jobard, B. & Sansom, M. S. P. (2014). Methodologies for the analysis of instantaneous lipid diffusion in MD simulations of large membrane systems. *Faraday Discuss.*, *169*, 1-18.

[25] Phillips, J. C., Braun, R., Wang, W., Gumbart, J., Tajkhorshid, E., Villa, E., Chipot, C., Skeel, R. D., Kal´e, L. & Schulten, K. (2005). Scalable molecular dynamics with NAMD. *J. Comput. Chem.*, *26*, 1781-1802.

[26] Shinoda, W., DeVane, R. & Klein, M. L. (2010). Zwitterionic lipid assemblies: Molecular dynamics studies of monolayers, bilayers, and vesicles using a new coarse grain force field. *J. Phys. Chem. B.*, *114*, 6836-6849.

[27] Ayton, G. S. & Voth, G. A. (2009). Hybrid coarse-graining approach for lipid bilayers at large length and time scales. *J. Phys. Chem. B.*, *113*, 4413-4424.

[28] Shelley, J. C., Shelley, M. Y., Reeder, R. C., Bandyopadhyay, S. & Klein, M. L. (2001). A coarse grain model for phospholipid simulations. *J. Phys. Chem. B.*, *105*, 4464-4470.

[29] Izvekov, S. & Voth, G. A. (2005). A multiscale coarse-graining method for biomolecular systems. *J. Phys. Chem. B.*, *109*, 2469-2473.

[30] Marrink, S. J., de Vries, A. H. & Mark, A. E. (2004). Coarse grained model for semiquantitative lipid simulations. *J. Phys. Chem. B.*, *108*, 750-760.

[31] Kabsch, W. & Sander, C. (1983). Dictionary of protein secondary structure: Pattern recognition of hydrogen-bonded and geometrical features. *Biopolymers*, *22*, 2577-2637.

[32] Rzepiela, A. J., Schafer, L. V., Goga, N., Risselada, H. J., de Vries, A. H. & Marrink, S. J. (2010). Reconstruction of atomistic details from coarse-grained structures. *J. Comput. Chem.*, *31*, 1333-1343.

[33] Lee, K. I., Jo, S., Rui, H., Egwolf, B., Roux, B., Pastor, R. W. & Im, W. (2012). Web interface for Brownian dynamics simulation of ion transport and its applications to beta-barrel pores. *J. Comput. Chem.*, *33*, 331-339.

[34] Delemotte, L., Tarek, M., Klein, M. L., Amaral, C. & Treptow, W. (2011). Intermediate states of the Kv1.2 voltage sensor from atomistic molecular dynamics simulations. *Proc. Natl. Acad. Sci. USA.*, *108*, 6109-6114.

[35] Delemotte, L., Dehez, F., Treptow, W. & Tarek, M. (2008). Modeling membranes under a transmembrane potential. *J. Phys. Chem. B.*, *112*, 5547–5550.

[36] Kutzner, C., Grubmüller, H., de Groot, B. L. & Zachariae, U. (2011). Computational electrophysiology: the molecular dynamics of ion channel permeation and selectivity in atomistic detail. *Biophys. J.*, *101*, 809-817

[37] Gumbart, J., Khalili-Araghi, M., Sotomayor, M. & Roux, B. (2012). Constant electric field simulations of the membrane potential illustrated with simple systems. *Biochim. Biophys. Acta.*, *1818*, 294-302.

[38] Huang, W. N., Sue, S. C., Wang, D. S., Wu, P. L. & Wu, W. G. (2003). Peripheral binding mode and penetration depth of cobra cardiotoxin on phospholipid membranes as studied by a combined FTIR and computer simulation approach. *Biochemistry*, *42*, 7457-7466.

[39] Wong, T. C. (2003). Membrane structure of the human immunodeficiency virus gp41 fusion peptide by molecular dynamics simulation II. The glycine mutants. *BBA-Biomembranes*, *1609*, 45-54.

[40] Bright, J. N. & Sansom M. S. P. (2003). The flexing/twirling helix: exploring the flexibility about molecular hinges formed by proline and glycine motifs in transmembrane helices. *J. Phys. Chem.*, *B*, *107*, 627-636.

[41] Petrache, H. I., Zuckerman, D. M., Sachs, J. N., Killian, J. A., Koeppe, R. E. & Woolf, T. B. (2002). Hydrophobic matching

mechanism investigated by molecular dynamics simulations. *Langmuir*, *18*, 1340-1351.

[42] Stockner, T., Ash, W. L., MacCallum, J. L. & Tieleman, D. P. (2004). Direct simulation of transmembrane helix association: role of asparagines. *Biophys. J.*, *87*, 1650-1656.

[43] Tieleman, D. P. & Sansom, M. S. P. (2001). Molecular dynamics simulations of antimicrobial peptides: from membrane binding to trans-membrane channels. *Int. J. Quantum Chem.*, *83*, 166-179.

[44] de Planque, M. R. & Killian, J. A. (2003). Protein–lipid interactions studied with designed transmembrane peptides: role of hydrophobic matching and interfacial anchoring. *Mol. Membr. Biol.*, *20*, 271-284.

[45] Li, Y. & Gong, H. (2015). Theoretical and simulation studies on voltage-gated sodium channels. *Protein Cell*, *6*, 413-422.

[46] Capener, C. E. & Sansom, M. S. P. (2002). Molecular dynamics simulations of a K+channel model: sensitivity to changes in ions, waters, and membrane environment. *J. Phys. Chem.*, *106*, 4543-4551.

[47] Delemotte, L., Treptow, W., Klein, M. L. & Tarek, M. (2010). Effect of sensor domain mutations on the properties of voltage-gated ion channels: molecular dynamics studies of the potassium channel Kv1.2. *Biophys. J.*, *99*, L72–L74.

[48] Khalili-Araghi, F., Tajkhorshid, E., Roux, B. & Schulten, K. (2012). Molecular dynamics investigation of the ω-current in the Kv1.2 voltage sensor domains. *Biophys. J.*, *102*, 258-267.

[49] Nishizawa, M. & Nishizawa, K. (2009). Coupling of S4 helix translocation and S6 gating analyzed by molecular dynamics simulations of mutated K_v channels. *Biophys. J.*, *97*, 90-100.

[50] Treptow, W., Marrink, S. J. & Tarek, M. (2008). Gating motions in voltage-gated potassium channels revealed by coarse-grained molecular dynamics simulations. *J. Phys. Chem.*, *B112*, 3277-3282.

[51] Crozier, P. S., Stevens, M. J., Forrest, L. R. & Woolf, T. B. (2003). Molecular dynamics simulations of dark-adapted rhodopsin in an explicit membrane bilayer: coupling between local retinal and larger scale conformational change. *J. Mol. Biol.*, *333*, 493-514.

[52] Govaerts, C., Blanpain, C., Deupi, X., Ballet, S., Ballesteros, J. A., Wodak, S. J., Vassart, G., Pardo, L. & Parmentier, M. (2001). The TXP motif in the second transmembrane helix of CCR5-structural determinant of chemokine-induced activation. *J. Biol. Chem.*, *276*, 13217-13225.

[53] Olivella, M., Deupi, X., Govaerts, C. & Pardo, L. (2002). Influence of the environment in the conformation of alpha-helices studied by protein database search and molecular dynamics simulations. *Biophys. J.*, *82*, 3207-3213.

[54] Kochva, U., Leonov, H. & Arkin, I. T. (2003). Modeling the structure of the respiratory syncytial virus small hydrophobic protein by silentmutation analysis of global searching molecular dynamics. *Protein Sci.*, *12*, 2668-2674.

[55] Kessel, A., Shental-Bechor, D., Haliloglu, T. & Ben-Tal, N. (2003). Interactions of hydrophobic peptides with lipid bilayers: Monte Carlo simulations with M2 delta. *Biophys. J.*, *85*, 3431-3444.

[56] Choma, C. T., Tieleman, D. P., Cregut, D., Serrano, L. & Berendsen, H. J. C. (2001). Towards the design and computational characterization of a membrane protein. *J. Mol. Graph. Model.*, *20*, 219-234.

[57] Fleishman, S. J. & Ben-Tal, N. (2002). A novel scoring function for predicting the conformations of tightly packed pairs of transmembrane alphahelices. *J. Mol. Biol.*, *321*, 363-378.

[58] Yu, C. H., Cukierman, S. & Pomes, R. (2003). Theoretical study of the structure and dynamic fluctuations of dioxolane-linked gramicidin channels. *Biophys. J.*, *84*, 816-831.

[59] Forrest, L. R., Tieleman, D. P. & Sansom, M. S. P. (1999). Defining the transmembrane helix of M2 protein from influenza A by molecular dynamics simulations in a lipid bilayer. *Biophys. J.*, *76*, 1886-1896.

[60] Yu, C. H. & Pomes, R. (2003). Functional dynamics of ion channels: modulation of proton movement by conformational switches. *J. Am. Chem. Soc.*, *125*, 13890-13894.

[61] Warshel, A. (2002). Molecular dynamics simulations of biological reactions. *Acc. Chem. Res.*, *35*, 385-395.

[62] de Groot, B. L., Frigato, T., Helms, V. & Grubmuller, H. (2003). The mechanism of proton exclusion in the aquaporin-1 water channel. *J. Mol. Biol.*, *333*, 279-293.

[63] Ilan, B., Tajkhorshid, E., Schulten, K. & Voth, G. A. (2004). The mechanism of proton exclusion in aquaporin channels. *Proteins*, *55*, 223-228.

[64] Phongphanphanee, S., Yoshida, N. & Hirata, F. (2008). On the Proton Exclusion of Aquaporins: A Statistical Mechanics Study. *J. Am. Chem. Soc.*, *130*, 5, 1540-1541.

[65] Tani, K., Mitsuma, T., Hiroaki, Y., Kamegawa, A., Nishikawa, K., Tanimura, Y. & Fujiyoshi, Y. (2009). Mechanism of Aquaporin-4's Fast and Highly Selective Water Conduction and Proton Exclusion. *J. Mol. Biol.*, *389*, 4, 694-706.

[66] Zhu, F. Q., Tajkhorshid, E. & Schulten, K. (2001). Molecular dynamics study of aquaporin-1 water channel in a lipid bilayer. *FEBS Lett.*, *504*, 212-218.

[67] Zhu, F. Q., Tajkhorshid, E. & Schulten, K. (2004). Theory and simulation of water permeation in Aquaporin-1. *Biophys. J.*, *86*, 50-57.

[68] Perozo, E. & Rees, D. C. (2003). Structure and mechanism in prokaryotic mechanosensitive channels. *Curr. Opin. Struct. Biol.*, *13*, 432-442.

[69] Anishkin, A., Gendel, V., Sharifi, N. A., Chiang, C. S., Shirinian, L., Guy, H. R. & Sukharev, S. (2003). On the conformation of the COOH-terminal Domain of the Large Mechanosensitive Channel MscL. *J. Gen. Physiol.*, *121*, 227-244.

[70] Bilston, L. E. & Mylvaganam, K. (2002). Molecular simulations of the large conductance mechanosensitive (MscL) channel under mechanical loading. *FEBS Lett.*, *512*, 185-190.

[71] Valadie, H., Lacapcre, J. J., Sanejouand, Y. H. & Etchebest, C. (2003). Dynamical properties of the MscL of Escherichia coli: a normal mode analysis. *J. Mol. Biol.*, *332*, 657-674.

[72] Perozo, E., Cortes, D. M., Sompornpisut, P., Kloda, A. & Martinac, B. (2002). Open channel structure of MscL and the gating mechanism of mechanosensitive channels. *Nature*, *418*, 942-948.

[73] Higgins, C. F. (1992). ABC Transporters-from Microorganisms to Man. *Annu. Rev. Cell Biol.*, *8*, 67-113.

[74] Hopfner, K. P., Karcher, A., Shin, D. S., Craig, L., Arthur, L. M., Carney, J. P. & Tainer, J. A. (2000). Structural biology of Rad50 ATPase: ATP driven conformational control in DNA double-strand break repair and the ABC-ATPase superfamily. *Cell*, *101*, 789-800.

[75] Chang, G. (2003). Structure of MsbA from Vibrio cholera: a multidrug resistance ABC transporter homolog in a closed conformation. *J. Mol. Biol.*, *330*, 419-430.

[76] Stenham, D. R., Campbell, J. D., Sansom, M. S., Higgins, C. F., Kerr, I. D. & Linton, K. J. (2003). An atomic detail model for the human ATP binding cassette transporter P-glycoprotein derived from disulfide cross-linking and homology modeling. *FASEB J.*, *17*, 2287-2289.

[77] Ma, J., Flynn, T. C., Cui, Q., Leslie, A. G., Walker, J. E. & Karplus, M. (2002). A dynamic analysis of the rotation mechanism for conformational change in F(1) -ATPase. *Structure*, *10*, 921-931.

[78] Melo, M. C. R., Bernardi, R. C., Rudack, T., Scheurer, M., Riplinger, C., Phillips, J. C., Maia, J. D. C., Rocha, G. B., Ribeiro, J. V., Stone, J. E., Neese, F., Schulten, K. & Luthey-Schulten, Z. (2018). NAMD goes quantum: an integrative suite for hybrid simulations. *Nature Methods*, *15*, 351-354.

[79] Schulz, G. E. (2002). The structure of bacterial outer membrane proteins. *BBA-Biomembranes*, *1565*, 308-317.

[80] Tamm, L. K., Abildgaard, F., Arora, A., Blad, H. & Bushweller, J. H. (2003). Structure, dynamics and function of the outer membrane protein A (OmpA) and influenza hemagglutinin fusion domain in detergent micelles by solutions NMR, *FEBS Lett.*, *555*, 139-143.

[81] Ferguson, A. D., Kodding, J., Walker, G., Bos, C., Coulton, J. W., Diederichs, K., Braun, V. & Welte, W. (2001). Active transport of an

antibiotic rifamycin derivative by the outer membrane protein FhuA. *Structure*, 9, 707-716.

[82] Monticelli, L., Robertson, K. M., MacCallum, J. L. & Tieleman, D. P. (2004). Computer simulation of the KvAP voltage-gated potassium channel: steered molecular dynamics of the voltage sensor. *FEBS Lett.*, 564, 325-332.

[83] Roy, N. K. (2019). A new semi-explicit atomistic molecular dynamics simulation method for membrane proteins. *Journal of Computational Methods in Sciences and Engineering*, 19, 259-286.

[84] Gedeon, P. C., Indarte, M., Surratt, C. K. & Madura, J. D. (2010). Molecular dynamics of leucine and dopamine transporter proteins in a model cell membrane lipid bilayer. *Proteins*, 78, 4, 797–811.

[85] Thomas, J. R., Gedeon, P. C., Grant, B. J. & Madura, J. D. (2012). LeuT Conformational Sampling Utilizing Accelerated Molecular Dynamics and Principal Component Analysis. *Biophys. J.*, 103, L1-L3.

[86] Wang, Y., Harrison, C., Schulten, K. & McCammon, J. A. (2011). Implementation of Accelerated Molecular Dynamics in NAMD. *Comp. Sci. Discov.*, 4, 015002.

[87] Shaw, D. E., Maragakis, P., Lindorff-Larsen, K., Piana, S., Dror, R. O., Eastwood, M. P., Bank, J. A., Jumper, J. M., Salmon, J. K., Shan, Y. & Wriggers, W. (2010). Atomic-Level Characterization of the Structural Dynamics of Proteins. *Science*, 330, 341-346.

[88] Kaback, H. R. (1992). The lactose permease of Escherichia coli: a paradigm for membrane transport proteins. *Biochim. Biophys. Acta*, 1101, 210-213.

[89] Vardy, E., Arkin, I. T., Gottschalk, K. E., Kaback, H. R. & Schuldiner, S. (2004). Structural conservation in the major facilitator superfamily as revealed by comparative modeling. *Protein Sci.*, 13, 1832-1840.

[90] Pang, A., Arinaminpathy, Y., Sansom, M. S. P. & Biggin, P. C. (2003). Interdomain dynamics and ligand binding: molecular dynamics simulations of glutamine binding protein. *FEBS Lett.*, 550, 168-174.

[91] Kimanius, D., Lindahl, E. & Andersson, M. (2018). Uptake dynamics in the Lactose permease (LacY). membrane protein transporter. *Scientific Reports*, *8*, 14324.

[92] Nikolaev, D. M., Shtyrov, A. A., Panov, M. S., Jamal, A., Chakchir, O. B., Kochemirovsky, V. A., Olivucci, M. & Ryantsev, M. N. (2018). A Comparative Study of Modern Homology Modeling Algorithms for Rhodopsin Structure Prediction. *ACS Omega*, *3*, 7555-7566.

[93] Lopes, P. E. M., Roux, B. & MacKerell, Jr. A. D. (2009). Molecular modeling and dynamics studies with explicit inclusion of electronic polarizability. Theory and applications. *Theor Chem Acc.*, *124*, 11-28.

[94] Meuwly, M. (2019). Reactive molecular dynamics: From small molecules to proteins. *WIREs Comput. Mol. Sci.*, *9*, e1386.

[95] Best, R. B. & Mittal, J. (2010). Protein Simulations with an Optimized Water Model: Cooperative Helix Formation and Temperature-Induced Unfolded State Collapse. *J. Phys. Chem. B.*, *114*, 46, 14916-14923.

[96] Brown, M. F. (2012). Curvature Forces in Membrane Lipid-Protein Interactions. *Biochemistry*, *51*, 9782-9795.

In: A Closer Look at Membrane Proteins ISBN: 978-1-53618-149-4
Editor: Tristan B. Møller © 2020 Nova Science Publishers, Inc.

Chapter 3

THE COMMANDMENTS OF STUDYING INTEGRAL MEMBRANE PROTEINS

Raymond J. Turner[*]
Department of Biological Sciences,
University of Calgary, Calgary, AB, Canada

ABSTRACT

As a budding Biochemist, I was introduced to Arthur Kornberg's ten commandments of enzymology. After 25 years of working in the field of integral membrane protein (IMP) structure-function, my trainees and students of my classes noticed that I would make statements on IMP research in the form of a commandment, in the guises of these early Biochemistry commandments. Here I share my commandments around IMP expression and purification, IMP biochemistry, IMP functionality studies, and IMP high-resolution structures.

---

[*] Corresponding Author's E-mail: turnerr@ucalgary.ca.

INTRODUCTION

A favorite read I give all my students on the introduction to the topic integral membrane protein (IMP) biochemistry; is the story of a novice membrane protein biochemist and how he learns to love Lysozyme (von Heijne 1999). This article is an informative yet playfully written review by Gunner von Hejine, where he good-humouredly highlighted the frustrations encapsulated in being a biochemist trying to study integral membrane proteins. Although considerable advancements in technology have occurred in the past 10 years since this review was written, IMPs are still notoriously more difficult to study than their soluble cousins.

This chapter aims to highlight some of the tripping points and mistakes to avoid or at least to be aware of, for those embarking down a path of study on an IMP. To deliver these thoughts in a highly consumable fashion; I will try to channel the great Arthur Kornberg's Ten Commandments of Enzymology (Kornberg 2000, 2003), which I also consider mandatory reading for every budding biochemist. Each commandment is stated and then explained in the context of understanding IMP structure-function.

THE COMMANDMENTS

Thou shalt...

Commandments Around IMP Expression and Purification

Not Strive to OVER-Express Integral Membrane Proteins (sic)

Well let's just say it, over-expressing IMPs can be frustrating at best. But let's correct this wording; what is really meant here is *Enhancing functional protein accumulation* and in this chapter, will refer to it as EFPA. Unfortunately, maximizing protein accumulation has been foolishly and incorrectly called "protein over-expression" for decades, leading to disgruntled comments from molecular biologists gleefully pointing out how dimwitted protein biochemists can be in their fundamental understanding of biology. This of course is not limited to IMP, but all EFPA. So, let's not continue to propagate this language. Proteins do not express, genes are expressed, subsequently producing mRNA, which is further translated into proteins. Thus, for EFPA to occur, we need several steps to be working at their optimal levels and properly coordinated. The process is far more complex than we typically consider, as not only does a protein have to be efficiently translated but in many cases, they are chaperoned in their folding, cofactors need to be added, they require cellular targeting as well as potentially assisted to assemble into a multiprotein complex. In the case of IMPs, 'properly targeted' means to the membrane (and the correct membrane in the cell), inserted into the membrane in the correct topology and post insertional folding and transmembrane segment assembly/interactions to occur (for a more complete description of these processes see review by Cymer et al. 2015). Furthermore, appropriate and crucial post-translational modifications of amino acids with methylations, glycosylation, lipidation, etc.

EFPA is notoriously challenging with IMPs, as the volume of protein capture in a cell (the membrane itself) is considerably more limiting than the cytoplasm, particularly when we consider Prokaryotic expression systems. EFPA of IMPs are typically plagued with problems with protein aggregations and inclusion bodies through the transmembrane helices (TMH) miss folding into beta-strands. Additionally, misfolded or slowly folding proteins will have exposed regions leading to degrees of proteolytic degradation (Schlegel et al. 2014).

Expression and translation at a high rate for some proteins can lead to high cellular stress. EFPA of a transporter can lead to uncoupling of the proton motive force (Winstone et al. 2002), or miss directed and non-specific transport across the membrane, both of which can lead to cellular stress or lethality. The stress of producing the IMP can actually lead to loss of expression through the selection of cells who are producing less IMP and thus more physiologically fit. This fitness selection can lead to plasmid curing and expression loss in yeast and bacterial expression systems.

Here it is also important to have some words about heterologous IMP expression. We observe that often IMP's do not have their codons optimized. The use of rare codons may be present to decrease the translation rate to coordinate more closely with the folding, targeting and insertion of the protein. Thus, methods to optimize codon usage for the expression of IMP can lead to folding and assembly problems and reduce IMP production (Schlegel et al. 2012). However, recently groups have gone after this issue exploring optimizing the transcription rates and codon usage which appears promising (see Claassens et al. 2017 and citations within for work in this area).

Overall, considerable time, effort, resources, and 'luck' must be put forward to optimize EFPA of IMPs as there are no effective rules of thumb. Of course, if going after a homolog or similar IMP family member, where previous work has provided a template toward EFPA, it is worth starting with such information and optimizing around it. But, for the most part, EFPA of IMPs is imperially determined. Thus, it should be considered (but is more often overlooked), that it is often more resource-efficient to accept the low initial yields one may start to get and simply do multiple preparations to get enough material for targeted experiments. Yet optimizing EFPA can be a wonderful learning experience for a graduate trainee, and gleefully fun for those that enjoy dancing with experimental conditions.

Not Forget about the Detergent

Throughout this chapter, I use the term 'membrane mimetic' rather than just detergent, as recently, alternative compounds have been explored to overcome issues with classical detergents. Regardless, most IMP researchers still use detergents (surfactants, amphiphiles, tensides, soaps) to solubilize their IMP of interest from their lipid home. To date, finding the best membrane mimetic is still mostly empirically determined, and various detergent companies now provide small samples in detergent array kits to explore detergent 'space' for your IMP of interest. Yet there are some things to consider. A key thing to remember is that one is trying to mimic the natural lipid environment as closely as possible.

Detergents are made up of essentially two parts. 1) The head, which is polar and can be anionic, cationic, zwitterionic, or non-ionic. 2) The tail, which is hydrophobic and can be flexible, straight, branched, or more ridged and steroid-like. Important considerations of working with detergents are overviewed in Tables 1 and 2.

Table 1. Important parameters of detergents

Parameter	What to think about
Critical micellar concentration (CMC)	Concentration of transition to begin forming micelles Dependent on T, pH, ionic strength.
Hydrophile-lipophile balance (HLB)	Expression of hydrophilic character HLB of 12-14.5 for IMP
Critical Micellular concentration (CMT)	also referred to Ghost point, as below CMT Detergents precipitate out into ghost white suspension.
Effect on biophysical parameters	Adds to molecular weight, Adds to the Hydrodynamic radius (Stokes radius) Influences migration on PAGE Can influence/inhibit ligand binding
Presence of aromatic rings	Can contribute to protein's Absorbance at 280nm

Table 2. Factors that affect detergent performance

Experimental factor	Effect
Temperature	CMT; working T,
pH	Detergent head group pKa; net charge
Ionic strength	Increase counter ion decreases CMC
Detergent concentration	CMC
Multivalent ions	Can cluster charged detergents and precipitate
Purity	Peroxides and aldehydes damage proteins Increase purity = increase price.
Organics	Decrease dielectric increase CMC
Protein concentration	Do you have enough detergent to solubilize your target protein

Alternatives to the use of classical detergents are still limited. Amphipols have been explored for some time, which are essentially polar or charged polymers decorated with acyl chains which are thought to wrap around the nonpolar domains of IMP providing aqueous solubility (Popot 2010). Attempts to use perfluorinated surfactants have shown some success as they can maintain IMP lipid contacts (Popot 2010). Another interesting group is the maltose neopentyl glycol amphiphiles, which gives two heads and two tails linked (Chae et al. 2010). The detergent alternative that has received the most hype recently are SMALPs. These are a Styrene Maleic Acid copolymer (SMA), which is considered to extract the IMP along with a small amount of the natural lipid bilayer encircled by the polymer to generate SMA Lipid Particles. SMALP scaffolded IMPs have quickly become popular with structural biochemists, particularly those using electron microscopy (Postis et al. 2015; Knowles et al. 2009).

Issues around the choice of detergent that are often overlooked is the IMP may have considerably different thermal and kinetic stability when solubilized in detergent compared to its natural lipid bilayer. Further, the detergent may not be a 'neutral' player and could potentially act as an inhibitor or allosteric regulator to an IMP activity.

To avoid issues around biochemical experiments in detergents, IMPs can be reconstituted into Lipid bilayer (typically of defined lipid content) vesicles to try to return to a natural state. Here one must consider the size of the vesicle and if membrane curvature could influence the biochemistry of the IMP of interest.

There are a few tricks that exist to avoid the use of detergents, which is to purify the IMP with its natural lipids. Along with SMALPs, Bicelles (Sanders and Prosser 1998) and Lipid nanodisks (Baybur & Sligar 2010) allow for natural lipid bilayer encapsulation but small enough for structural studies. An approach to avoid detergent and lipid completely has been introduced by expressing the IMP as a protein-peptide fusion construct referred to as the SIMPLEx system (Mizrachi et al. 2015). The IMP remains soluble and folded correctly with the amphipathic peptide providing a hydrophobic shield around the IMP so that the in a protein.

Regardless of what membrane mimetic one chooses one needs to remember it is NOT the natural environment of the IMP and changes to dynamics and stability *will* occur. Even with the lipid-based systems of SMALPs, bicelles, and nanodisks, the lateral pressure and diffusability will be different, and of course, the topological sidedness of the protein will be lost. Additionally, one should consider that detergents can have deleterious interactions with the extramembranous soluble domains (Yang et al. 2014). It is fundamental to remember that a solubilized IMP is a fish out of water, still a fish, but its behaviour can be very different.

Not Overlook the Additional Challenges of IMP Purification

For IMPs the first challenge is optimizing EFPA, but one typically wants to purify the protein for structure-functional studies. For the most part, one can follow the plethora of protein purification approaches used for soluble proteins (see Deutscher 1990 as a classical excellent resource). As with soluble proteins, one must concern oneself about the variables of cell lysis approaches and the protection of protein from proteases and oxidative stressors. Yet, the choice of temperature, buffers, and metal/salt ions and concentrations becomes more complicated due to the need to use membrane mimetics to solubilize the protein (see other commandments as well).

The factors highlighted in Tables 1 and 2 can lead to serious issues in protein purification as the net charge of the protein is influenced by the charges of the detergent and the size of the protein is increased with the detergent shell. Thus, chromatographic purification approaches behave considerably different as well as how the protein will behave when centrifuged. A common error in purifying a detergent-solubilized IMP with ion-exchange chromatography is forgetting the net charge of the head group of the detergent and only focusing on the calculated pI of the target protein. The reader is directed to sections I and II of Hunte et al. (2003), for examples and more detail on each purification method's considerations.

An extremely popular approach to facilitate protein purification is to add an affinity tag through molecular biology approaches. The most frequently used tag is that of the hexa-histidine (His_6) peptide added for use on immobilized metal ion absorption/affinity chromatography (IMAC). Undoubtedly, affinity-affinity tags (Terpe 2003) have been the most significant technological advancement helping accelerate protein biochemistry. However, there are issues with such tags influencing and changing the activity and stability of the tagged target protein (Booth et al. 2018; Majorek et al. 2014; Mohanty & Weiner 2014). An example of the influence of a His_6 tag on an IMP transporter is seen with the bacterial multidrug resistance transporter EmrE where the *in vivo* resistance profile was changed as well as changes of *in vitro* structural observations regarding multimeric state equilibrium and ligand binding upon the addition of the tag (Qazi et al. 2015). Therefore, one can still utilize affinity tag approaches to initiate the purification, but do not consider that your protein will behave as wildtype *in vivo* or reconstituted and that the tag may lead to miss-interpretations in structural and functional studies.

Commandments Around IMP Biochemistry

Understand the Differences in Stability Compared to Globular Proteins

IMPs are remarkably thermodynamically stable in the lipid bilayer. This comes from the hydrogen bonds fully satisfied in the hydrophobic environment where the weakly coulombic nature of the H-bond is enhanced in the hydrophobic environment. Similarly, salt bridges are stronger and the intimate contacts of the van der Waals interactions between helicies provide large enthalpic contributions in addition to the balance of the lipophobic effect and hydrophobic matching towards entropy contributions (Engelman et al., 2003; White & Wimley 1999; Bowie 2005).

The mixed solvation conditions (lipid environment and bulk aqueous phase) leads to unexpected behavior in traditional biochemical protein manipulation conditions. Consider denaturation using Urea or Guanidinium ion. Both of these denature a protein by competing out H-bonds. In the case of an IMP in a lipid or membrane mimetic, the extra membrane domains will denature but the transmembrane polypeptide remains inaccessible to these compounds and does not denature. Thermal denaturation is different as well due to the large enthalpy and many IMPs do not thermally denature. If they do, upon cooling they often aggregate into beta-strand based amyloid structures. Certainly, the delipidated protein is highly unstable and why IMPs are manipulated in membrane mimetics.

A traditional storage approach of proteins is Lyophilizing, or -20°C storage. Lyophilisation can concentrate salt ions leading to aggregation events with detergents. The freezing step can trap crystallization water within the detergent protein complex dissociating TMH leading to protein aggregation. As with any protein storage conditions need to be conducted.

Many small IMPs with short extra membrane loops do not denature (linearize) in sodium dodecyl sulfate (SDS). In fact they can maintain their structure and ligand binding properties in such detergents (Tulumello & Deber, 2012). The use of this detergent in electrophoresis techniques is standard biochemistry and migration comparisons are based on the ratio of SDS molecules with molecular weight to define a charge to mass ratio. For IMPs, they can bind differential amounts of SDS (depending on the degree of TMH-TMH interactions are disrupted) leading to anomalous migration on SDS-PAGE (Rath et al. 2009). There is also the issue of the competitive ability of SDS to outcompete the membrane mimetic used for the solubilization steps.

Tips to deal with these issues experimentally are provided in Hunte et al. (2003), but overall it is important to recognize stability differences compared to globular proteins.

Think Carefully of Experimental Conditions

Biochemists will think carefully of their experimental physicochemical conditions: pH and buffering compound, ionic strength, counterions, dielectric constant, redox potential, temperature, and protein concentration/density.

These conditions are chosen with their protein of interest in mind to try to mimic the natural physiological conditions as well as stabilize their system while doing experiments. In this regard, we work in a biological buffer with a mixture of various compounds and concentrations to define the ideal condition.

However, when working with IMPs we have an additional compound, the membrane mimetic, which is either influenced or influences the other biological buffer components. If we don't want our detergent, and subsequently our IMP, precipitating out we need to consider how the biological buffer components will influence the detergent performance (consider Tables 1 and 2).

Temperature is likely the most frequent mistake of the novice IMP biochemist in the lab. This comes from the standard of doing all protein experiments on ice or in the cold room. Yet, many detergents have their CMT between 4 and 24°C, which leads to the detergents crystalizing and precipitating out or forming gel-like phases.

Another consideration is in moving from the detergent or detergent-like compounds being used as membrane mimetics to reconstitution into proteoliposomes. Considering first that reconstitution is an art form in itself and this process is often different for each unique IMP. Then one must consider the lipid composition of the liposome as well as the size. Will curvature strain be an issue for ones given IMP; ie. does it originate from a membrane with high curvature or more planar? Or more fundamentally, are your chosen lipids appropriately hydrophobically matched (Killian 1998)? Even considering the head group and lipid mixture is of importance regarding maintaining topology. Biological lipids have different lipid compositions in each leaflet of the bilayer and protein sequences have evolved to match.

It is extremely difficult, if not impossible to mimic accurately all the biological conditions. So beyond knowing this as a biochemist, here I leave the warning to be extremely careful about the interpretation from your model system to what is going on in the cell.

Commandments Around IMP Functionality Studies

Not Ignore Ligand Binding Differences

Kinetics and diffusion in 3D space are considerably different than the 2D plan. Here we consider the differences in the ability to bind ligand. The ability of a substrate to find the binding site of an IMP transporter or receptor in a lipid bilayer will be different than when it is solubilized in a membrane mimetic allowing extra degrees of freedom. This will affect the 'on rates' and subsequently the binding constant. Differences in this diffusion property on the kinetics are complemented by differences in thermodynamic energy properties of ligand-protein binding. The dynamics of the peptide chain and subsequent movements of amino acid side chains would be modulated differently by the membrane mimetic compared to the lipid bilayer.

Thus the on/off rates can be different leading to very different binding constants. An early study evaluating Nucleoside transporters noted that different detergents needed to be screened to find one that got close to retaining the high ligand affinity observed *in vivo* (Hammond & Zarenda. 1996). A study illustrating these differences of membrane mimetic on structure and dynamics, circular dichroism and fluorescence were performed on the model multi-ligand transporter, EmrE. Remarkable differences were observed with the protein solubilized in different detergents as well as vesicle systems and solvents (Federkeil et al. 2003). A follow-up study using isothermal titration calorimetry confirmed the above statements with K_d for the same substrate ranging over tenfold (Sikora & Turner, 2005). A more recent example is seen with P-glycoprotein where modulators stimulated ATPase activity compared to inhibition in native membranes (Shukla et al. 2017).

The issues around detergent selection for maintaining activity have been recognized since the 1960s with studies of different detergents on mitochondrial enzymes (example see Soltysiak & Kaniuga 1970). Unfortunately, such issues became forgotten with high throughput purification methods, as there are very few studies evaluating changes to the structure, binding with a change of membrane mimetics. Yet this issue is found in various early texts on membrane proteins (early attention to this issue brought up in Gennis. 1989). Few would do activity comparisons under different solubilization conditions as, well, why would they? If after a frustrating timeframe optimizing EFPA and detergent selection and purification, finally finding something that worked, few would want to start again.

So, this commandment is simply to note that there is a strong likelihood that different solubilization and physicochemical conditions will give different answers about your IMP of interest.

Remember That Membrane Sidedness Is Lost

This seems obvious, but it tends to be forgotten or ignored. Upon solubilized of an IMP in a membrane mimetic, it is important to recognize that the biological sidedness of the protein has now been lost. The biological membrane is referred to in most textbooks as a semi-permeable barrier and divides different environments. Consider the protein in Figure 1. Each side of the membrane has completely different physicochemical environments. There can be remarkable differences in pH, ionic strength, specific ions, and ion concentrations, dielectric constant, redox potential, and both low and high molecular weight metabolites and biomolecules (peptides, proteins, carbohydrates, nucleic acids). There will also be differences in fluidity in the lipid leaflet on each side of the membrane with conditions of side influencing the protein dynamics, ligand binding, and catalytic activity.

Biologically, an IMP has evolved so that the amino acid sequence on the different sides of the membrane is well suited to those specific physicochemical conditions. Thus, once the protein is solubilized, the extra membrane domains are now exposed to identical conditions, defined by the experimentalist, which may or may not be relevant to either side of the membrane. Replacing detergent membrane mimetics with bilayer disks can solve the issues around the accuracy of the detergent to replicate the bilayer, yet the sidedness is still lost leaving the extra membrane domains exposed to the same conditions.

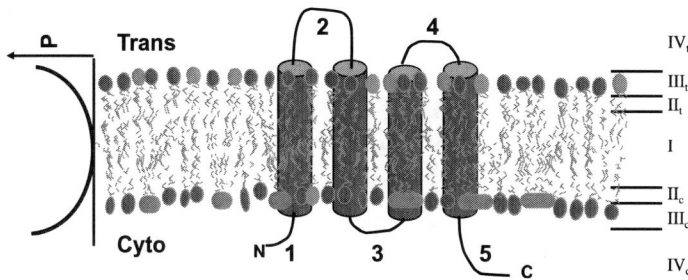

Figure 1. Depiction of biomembrane with a protein topology sketch. Cartooned here is a bilayer demonstrating different lipid composition between the two leaflets of the membrane. The two sides of the membrane are defined as the cytoplasmic side or cis side (subscript 'c') and the outside or trans side (subscript 't'). On the right in roman numerals indicate the different regions of the membrane with region: I as the acyl chain lipid core; II as the slightly polar glycerol region of the glycerol lipids; III as the highly polar and or charged lipid head group region potentially also influenced by a phosphoester bridging group; IV the bulk solvent. The subscripts of these regions are to depict that their physicochemical environments are different to each other. Considering the potential of differential saturation of acyl chains, once could also divide region I to the c and t leaflets as the dynamics and packing pressure can also be different. The graph depiction on the left illustrates the nature of the hydrophobic (polarity; P) gradient that exists, and that it is not an instant transition. A topology model of a 4 transmembrane helix protein is also shown with the numbers indicating the different extra membrane loops. From this diagram one should recognize that the chemical environments of loops 1, 3, and 5 would be different from loops 2 and 4.

Not Confuse In Vitro vs In Vivo Activity

I have been amused over the years receiving reviewer comments asking for activity measurements of the detergent-solubilized transporters I have worked on. Given that a transporter moves a ligand from one side of the membrane to the other, such activity by definition is lost once solubilized. A similar issue exists for ion channels and receptors. This is the ramification of the biomembrane sidedness and solubilization causing forfeiture of physicochemical condition separation as in the commandments and comments above. However, one can still measure ligand binding as a proxy, but the binding observed will likely be somewhat disconnected to the *in vivo* state. Not to say such experiments are not worth doing to compare mutants, ligand specificity, and inhibitors.

For the most part, it is still impossible to obtain biochemical structure-function information of an IMP while in its natural cellular environment *in vivo*. Certainly, experiments in proteoliposomes, defined detergent vesicle, or black membrane systems provide closer to *in vivo* realities, but they may still not be able to provide exactly replicated conditions. Yet combining genetic phenotype experiments beside good cell biology, complemented with *in vitro* biochemical and structural studies a considerable wealth of knowledge has been amassed on IMPs. The message here is simply to recognize that our experimental model systems still do not precisely mimic those of a cell.

Commandments Around IMP Structure

Not Put All Faith in Hydropathy Plots

I still find myself shocked when in a graduate student committee meeting where they are announcing they have cloned a gene responsible for this or that, and they had performed bioinformatic analysis and go on to define it as an IMP and show me a picture of the protein winding back and forth through a double-lined membrane followed by excitement around some domain of residues as binding site or the like. Such meetings remind me of the blind faith students (and many senior researchers) often have in some of our tools, using them as black boxes. But in the case of IMPs, not understanding the issues around hydropathy analysis and various prediction algorithms frequently leads to wrong models and subsequent downstream problems in experimental interpretations. Although many of the present programs work remarkably well, however, they may never be 100% accurate.

The first comment on this issue is the choice of hydropathy scale that is applied to the primary amino acid sequence. For the uninitiated, this first step is considered trivial. Assign a value of hydrophobicity to each amino acid and apply these values to the primary sequence as initially applied by Kyte & Doolittle (1982). However, it is remarkably challenging to agree on such a value for each amino acid. More than 80 hydrophobicity/hydrophilicity indices existed by 1989, and the next two decades saw on the order of 2-4 new hydropathy scales proposed per year. Some became favored for various reasons, but often would be chosen by default or random and applied to a sequence to generate hydropathy plots predicting TMH regions, which were taken on faith to be accurate, but overall were disappointingly error-prone (Elofsson & von Heijne, 2007).

Advancements came as the field recognized that some classes of IMPs were more efficiently predicted with some scales versus others (Crimi & degli Esposti, 1991; Turner & Weiner, 1993). Further improvements came through considering the ionized versus a neutral form of amino acid, as well as considering possible charge pairing would neutralizing charges for membrane insertion (Jayasinghe et al. 2001). Another advancement was to recognize the regions in the membrane (Figure 1) and that the hydropathy values should not be weighted equally to the core of the membrane versus the membrane barrier as there is a polarity gradient in the natural membrane (von Heijne & Rees 2008).

Issues with hydropathy analysis are nicely illustrated by applying different scales to the same protein, which will often lead to different outcomes. With applying the learned concepts the scale outputs are getting closer to an agreement. However, using different approaches still leads to a differences in the confidence of the final predictions (MacCallum & Tieleman 2011).

Using hydropathy knowledge in combination with homology sequence alignments and computational learning increased the predictability of multispanning IMPs to be over 80% accurate (Viklund & Elofsson 2004). However, the false positive and false negative rates (either predicting a region to be in the membrane or missing a transmembrane region) for most algorithms are still on the order of 20-30% (Zhao et al. 2006). An explanation given for why we may not be able to reach 100% accuracy, is that some TMH could be fully hidden from the hydrophobic lipids and thus 'look like' a soluble helix. Further, some sequences in a large soluble protein may be sufficiently hydrophobic, being they are hidden from the polar aqueous environment by the remainder of the protein (Zhao et al. 2006). Thus, some TMH may be impossible to distinguish from helicies in soluble globular proteins and *visa versa*. Furthermore, even with the significant improvements of prediction algorithms, it is still very difficult to predict marginally hydrophobic TMH, disrupted TMH, and re-entrant loops such as what seems to occur in the glutamate transporter and some ion channels.

Even with this knowledge, taking a quick look online gives the observation that there are still a large number of servers that default to a single Kyte and Doolittle scale and simple calculations. A list of good predictor algorithms was put together by Tsirigos et al. (2013). My recommendation here is to use different algorithms, choosing different hydropathy scales and selectivity limits and take the average result as your working model.

Not Believe Blindly the Predicted Topology

Moving from the prediction of the TMH by hydropathy analysis, one normally takes the subsequent leap to define the topology of the protein in the membrane; the winding of the peptide sequence back and forth across the membrane defining the location of specific extra membrane loops/domains and location of the N and C-termini. This was initially considered a trivial exercise as one should be able to define if the protein has a cleavable signal sequence or not, and then just wind the protein back and forth across the membrane from that start position based on hydropathy analysis. However, it is now recognized that IMP biogenesis can follow many pathways for insertion, folding and maturation (White 2015; Cymer et al. 2015), complicating this early assumption.

An early improvement to hydropathy analysis was the recognition that prokaryotic proteins had a sidedness preference of charged amino acids allowing one to test one's helix number predictions to the positive-inside rule (von Heijne, 1992). However, a complication to this rule has come from recognizing that lipid type also influences the topology (Dowhan & Bogdanov, 2009; Dowhan et al., 2019). A further cautionary note recognizes that protein topology can vary between members of the same functional family (Tsirigos et al. 2018).

The Commandments of Studying Integral Membrane Proteins 137

There are a wide range of sidedness fusion tags and probes for both eukaryotic and prokaryotic IMP to experimentally test the predicted topology (van Geest & Lolkema, 2000). Additionally, a curated database of IMP topologies, MPtopo (initially beginning from Jayasinghe 2001b) now exists, which is a value of this database is that it is conservative, containing only experimentally validated transmembrane segments.

Overall, the two-dimensional topology model is the first step to move forward for structure-functional biochemistry. Having the topology right early on helps guide experiments towards a biological mechanism. If one misinterprets the topology, one will be plagued with confusing results moving forward.

Be Skeptical of High-Resolution Structures

There are a number of studies that point out the gap between the number of sequences representing IMP in sequenced genomes – typically on the order of 30% - and the low number of known unique IMP structures (see https://blanco.biomol.uci.edu/mpstruc). This differential illuminates the challenges along the structure determination path for IMPs.

A degree of cautionary thought of a structure's accuracy should be considered for all structural biology, as in the case of crystallography one is growing crystals outside the *in vivo* physicochemical conditions. It is often forgotten that the final structure is a *model* to represent the data. Considering when it comes to IMPs solved in a membrane mimetic, one's skepticism should be heightened. This is not to say that the 3D structures are wrong. Certainly, the polypeptide folded and interacted in a way that promoted an ordered crystal and this growth was under the conditions of the experimentalist screened to find the ideal conditions for this growth. But these final conditions are rarely what one would find *in vivo*; nature evolved not to crystalize their proteins for the most part. So, with such aggressive processing of the sample and the difficulty of mimicking the natural conditions, one should not be surprised that artifacts slip in. The challenge is understanding the relevance of the given structure to the *in vivo* state, and in the context, here, in a lipid bilayer of a specific membrane/cellular environment. Additionally, heterologous protein-

protein interactions are typically lost or ignored as one is crystalizing the minimal unit (as with the majority of studies in the past). It is extremely difficult to mimic the conditions *in vivo* for a soluble protein, and exceedingly more complicated for an IMP where there are multiple physicochemical regions as lamented on in other commandments.

In earlier times, prior to about 2015, we did not appreciate the role of lipids in modulating the structure of IMP, thus most deposited structures solved by X-ray crystallography methods used IMP solubilized in detergents. The lipid was considered a contaminant and if found as part of the structure the molecules were considered an experimental artifact in the same manner of a buffer molecule found tightly bound in a soluble protein structure. This led to several structures deposited, that although were solved correctly, did not represent the correct biologically functional lipid constrained structure. If one accepts that any given type of IMP will have a unique annular lipid and specific lipid-protein interactions (Corradi et al. 2019). Using different membrane mimetics for a given IMP should lead to at least subtle variations in the resulting structures. Similarly, using the same membrane mimetic for different proteins one should expect variations of influence on the different structures. This, of course, has led to the frustrations between different groups' data and interpretations.

It was also remarkable to see structures discussed with great enthusiasm towards mechanism interpretation, even though the unique detergent-solubilized-crystalized form did not correspond to the biochemistry performed in other membrane mimetics. The worst cases of these involved early structures of the transporters P-glycoprotein and EmrE where structures needed to be retracted (although I note they are still accessible). Different membrane mimetics can lead to the stabilization of different structural states over others. An example is seen with the Ryanodine Receptor (Willegems and Efremov 2018). This recent and many early studies show the importance of caution in interpretation, without which has led to biases and assumptions on IMP structures and subsequent interpretations in function.

Unfortunately, little work has been done to decipher the mechanisms that are influencing the structure of an IMP in protein-detergent complexes and the effects of different folding states. In the past decade, more researchers are appreciating that the choice of detergent has to go beyond getting a good extraction and protein purification yield. It is no longer acceptable to ignore that a chosen membrane mimetic may not retain the correct structural and physical properties of the protein. (Anandan & Vrielink 2016). The modulations may be as subtle as a TMH rotated away from its native contacts, the TMH kinked differently, or even to the extent a TMH is no longer defined in the bilayer.

Related to this it is worthwhile to remember that up to about 2010, it was dogma that the transmembrane structure of an IMP was either all beta or all alpha helix. Albite, it was recognized that different pitches could occur within a given helix. It was also considered that once the polypeptide strand penetrated the lipid bilayer it would continue all the way through as a transmembrane segment (either helix or beta). With this bias of interpretation in mind, if a structure was solved with transmembrane segment abnormalities, the structure was typically thrown out. Since then structures of mixed secondary structure within the membrane have been solved, and several with interrupted secondary structure and those with an incomplete transmembrane strand turning to return to the same side. Another striking failure of earlier structures was defining membrane sidedness to a detergent-solubilized structure in *ab initio* of other biochemical data.

Combining the advances in cryo-electron microscopy with lipid disk membrane mimetic approaches leads us into a new revolution in IMP structural biology. The warning here is to be careful not to get seduced by the hype and continue to recognize potential experimental artifacts creeping into structures to mislead mechanistic interpretations.

Remember the Importance of the Lipid

A theme through many of these commandments is the solvent of IMPs is a multiphasic mixture of ionic water and lipids. There is considerable support now to appreciate the role of lipids in IMP folding. We also now appreciate the regions of a membrane better which includes: the core where the acyl chains reside, the glycerol region of the lipid, and the head group region (Figure 1, regions I, II, II). The dynamics and dielectric constant of these regions are quite different as well as the ability of water to penetrate which leads to preferences of where amino acids prefer to sit in the membrane. Additionally, the two leaflets of the lipid bilayer can be remarkably different in lipid content leading to different thicknesses and bilayer dynamics (Ingolfsson et al. 2014). The differences in lipid composition can also flip the topology of an IMP (Bay & Turner, 2013). A more detailed discussion of lipids and IMP see reviews by Corradi et al. (2018; 2019).

Serious improvements in solubilisation methods and novel membrane mimetics have been developed to allow the reconstitution of IMPs into a lipid disk environments suitable for structure evaluation by cryo-EM, NMR, SAXS, neutron scattering and using cubic lipids for crystallography. These systems have allowed biophysical chemists to overcome earlier limitations. Also, now structures are complemented with computational modeling and molecular dynamics has led towards increased faith of the resulting structure as well as noting the lipid environment in more detail.

Yet, still, the majority of the biochemical and structural work is done with the IMP purified and solubilized in membrane mimetic environments, particularly detergents. Regardless, with the advancement of computational approaches, structures solved in detergent can be placed into a lipid bilayer *in silico* and molecular dynamics simulations performed to 'solve' a more biologically relevant structure. Now this task is less difficult the group of Sansom produced a platform to facilitate putting IMP into lipid membranes - MemProtMD (Stansfeld et al. 2015).

CONCLUSION

In no means do I consider to hold a candle to the protein prowess of G. Von Heijne and A. Kornberg for their experience with proteins that lead to their informative prose (von Heijne 1999, Kornberg 2003). Regardless, the above commandments come from my >25 years of biochemical experimental studies on IMP structure function and all the Love and Hate that such an experience brings. Some of these commandments relate to issues around our early naivety in the field's perceived understanding of the influence that various experimental parameters have on IMP structure and function. It is impossible to know how much *dogmatic* information we have accumulated on specific IMPs that could be unwittingly infected by breaks in these commandments leading to artifacts awash in the literature. This chapter was not meant to be an extensive review of each of the issues highlighted, only to lead the novice membrane protein biochemist towards finding the right path(s). IMPs are arguably the most import group of proteins, coding at least a third of the genomes (no bias here, *honest*) and thus drive the adventurous and courageous towards lifelong relationships. Therefore, I leave these commandments to the IMP research field to consider for both moving forward as well as when contemplating the literature of the past.

REFERENCES

Anandan, A. Vrielink, A. 2016. Detergents in membrane protein purification and crystallization. In The Next generation in Membrane protein structure determination. *Adv. Exp. Med. Biol.* 922:13-28.

Bay, DC. Turner RJ. 2013. Membrane composition influences the topology bias of bacterial integral membrane proteins. *BBA-Biomembranes.* 1828: 260-270.

Bayburt, TH. Sligar, SG. 2010. Membrane protein assembly into nanodiscs. *FEBS Letters* 584:1721-1727.

Booth, WT. Schlachter, CR. Pote, S. Ussin, N. Mank, NJ. Klapper V. Offermann, LR. Tang, C. Hurlbeut BK, Chruszcz, M. 2018. Impact of an N-terminal Polyhistidine tag on protein thermal stability. *ACS Omega* 3: 760-768.

Bowie, JU. 2005. Solving the membrane protein folding problem. *Nature* 438:581-589.

Chae, PS. Rasmussen, RR. Rana et al. 2010 Maltose neopentyl glycol (MNG) amphiphiles for solubilization, stabilization and crystallization of membrane proteins. *Nature Methods* 7:1003-1008.

Claassens, NJ. Siliakus, MF. Spaans, SK. Creutzburg, SCA. Nijsse, B. Schaap, PJ. Quax, TEF. Van der Oost, J. 2017. Improving heterologous membrane protein production in Escherichia coli by combining transcriptional tuning and codon usage algorithms. *PLoS ONE* 12: e0184355.

Corradi, V. Sejdiu BI, Mesa-Galloso H. Abdizadeh H. Noskov S. Marrink SJ. Tieleman DP. 2019. Emerging diversity in lipid-protein interactions. *Chemical Reviews*. 119:5775-5848.

Corradi, V. Mendez-Villuendas, E. Ingolfsson, HI. Gu, R-X. Siuda I., Melo, MN. Moussatova, A. DeGagne LJ, Sejdiu BI. Singh G., Wassenaar, TA, Magnero D. Marrink SJ. Tieleman DP. 2018. Lipid-Protein interactions are unique fingerprints for membrane proteins. *ACS Central Sciences*. 4: 709-717.

Crimi, M. Degli Esposti, M. 1991. Structural predictions for membrane proteins: the dilemma of hydrophobicity scales. *Trends Biochemical Science* 16:119.

Cymer, F., von Heijne G., White, SH. 2015. Mechanisms of integral membrane protein insertion and folding. *J. Molecular Biology*. 13:999-1022.

Deutscher, MP. 1990. Guide to Protein Purification. *Methods in Enzymology, Volume 182*. Academic Press.

Dowhan W, Bogdanov M. 2009. Lipid-dependent membrane protein topogenesis. *Annu Rev Biochem*. 78:515-40.

Dowhan W, Vitrac H. Bogdanov M. 2019. Lipid-assisted membrane protein folding and topogenesis. *Protein J*. 38:274-288.

Elofsson A, von Heijne G. Membrane protein structure: Prediction versus reality. *Annu Rev Biochem*. 2007;76:125–140.

Engelman, DM. Chen Y. Chin C-N. Curran AR. Dixon AM. Dupuy AD. Lee AS. Lehner U. Matthews EE. Reshetnyak YK. Senes A. Popot J-L. 2003. Membrane protein folding: beyond the two-stage model. *FEBS Letters* 555; 122-125.

Federkeil, S. Winstone, TL, Jicking G. Turner, RJ. 2003. Spectroscopic Analysis of EmrE in Various Membrane Mimicking Environments. *Biochemistry Cell Biology* 81:61-70.

Gennis, RB. 1989. *Biomembranes, molecular structure and function.* Springer-Verlag.

Hammond, JR. Zarenda, M. 1996. Effect of detergents on ligand binding and translocation activities of solubilized/reconstituted nucleoside transporters. *Arch. Biochemistry Biophysics.* 332:313-322.

Hunte, C. von Jagow, G. Schagger, H. 2003. *Membrane protein purification and crystallization. A practical Guide.* Academic Press.

Ingolfsson, HI. Melo, MN. van Eerden, FJ. Arnarez, C. Lopez, CA. Wassenaar, TA. Periole X. de Vries, AH. Tieleman, DP, Marrink SJ. 1014. Lipid organization of the plasma membrane. *J. Am. Chem. Soc.* 136: 14554-14559.

Jayasinghe, S. Hristova, K. White, SH. 2001. Transmembrane helix energetics and prediction accuracy. *J. Molecular Biology* 312:927-934.

Jayasinghe, S. Hristova, K. White, SH. 2001b. MPtopo: A database of membrane protein topology. *Protein Sciences* 10:455-458.

Killian JA. 1998. Hydrophobic mismatch between proteins and lipids in membranes. *Biochimica Biophysica Acta – Reviews on Biomembranes. 1376*:401-416.

Knowles TJ. Finka R. Smith C. Lin YP. Dafforn T. Overduin M. 2009. Membrane proteins solubilized intact in lipid containing nanoparticles bounded by styrene maleic acid copolymer. *J. Am. Chem. Soc.* 131:7484–7485

Kornberg, A. 2000. Ten commandments: Lessons from the Enzymology of DNA Replication. *J. Bacteriology.* 182:3613-3618.

Kornberg, A. 2003. Ten commandments of enzymology, amended. *Trends Biochemical Sciences.* 28: 515-517.

Kyte J, Doolittle RF. A simple method for displaying the hydropathic character of a protein. *J Mol Biol.* 1982;157:105–132.

MacCallum, JL. Tieleman, DP. 2011. Hydrophobicity scales: a thermodynamic looking glass into lipid-protein interactions. *Trends Biochemical Sciences* 36:653-662.

Majorek, KA. Kuhn, ML. Chruszcz, M. Anderson, WF. Minor, W. 2014. Double trouble - Buffer selection and His-tag presence may be responsible for non-reproducibility of biomedical experiments. *Protein Science.* 23:1359–1368.

Mizrachi, D. Chen, Y. Ke, A. Pollack, L. Turner, J. Auchus, RJ. DeLisa MP. 2015. Making water-soluble integral membrane proteins in vivo using an amphipathic protein fusion strategy. *Nature Communications* 6:6826

Mohanty, AK. Wiener, MC. 2004. Membrane protein expression and production: effects of polyhistidine tag length and position. *Protein Expression Purification* 33:311–325.

Popot, J.-L. 2010. Amphipols, nanodiscs and fluorinated surfactants: three non-conventional approaches to studying membrane proteins in aqueous solutions. *Annual Rev. Biochemistry* 79: 737-775.

Postis, V. Rawson, S. Mitchell, JK. Lee, SC, Parslow, RA, Daffon, TR, Baldwin SA. Muench SP. 2015. The use of SMALPs as a novel membrane protein scaffold for structure study by negative stain electron microscopy. *Biochim Biophys Acta* 1848; 496-501.

Qazi, SJS. Chew, R. Bay DC, Turner RJ. 2015. Structural and functional comparison of hexahistidine tagged and untagged forms of small multidrug resistance protein, EmrE. *Biochemistry Biophysics Reports* 1;22-32.

Rath, A. Glibowicka M. Nadeau VG. Chen G. Deber CM. 2009. Detergent binding explains anomalous SDS-PAGE migration of membrane proteins. *PNAS* 106:1760-1765.

Sanders, CR. Prosser, RS. 1998. Bicelles: a model membrane system for all seasons? *Structure* 6: 1227-1234.

Schlegel S, Lofblom J, Lee C, Hjelm A, Klepsch M, Strous M, Drew, D. Slotboom DJ. De Gier, J-W. 2012 Optimizing membrane protein overexpression in the *Escherichia coli* strain Lemo21(DE3). *J Molecular Biology* 423: 648–59.

Schlegel S, Hjelm A, Baumgarten T, Vikstrom D, de Gier J-W. 2014. Bacterial-based membrane protein production. *Biochim Biophys Acta.* 1843: 1739–1749.

Shukla, S. Abel, B. Chufan EE., Ambudkar, SV. 2017. Effects of a detergent micelle environment on P-glycoprotein (ABCB1)-ligand interactions. *J. Biological Chemistry.* 292:7066-7076.

Sikora, CW., Turner, RJ. 2005. Investigation of Ligand binding to EmrE by Isothermal Calorimetry. *Biophysical J.* 88: 475-482.

Soltysiakm D. Kaniuga Z. 1970. The effect of Triton S-100 on the respiratory chain enzymes of a heart-muscle preparation. *FEBS J.* 14: 70-74.

Stansfeld, PJ. Goose, JR. Caffrey M. Carpenter EP." Parker JL. Newstead S. Sansom MS. 2015. MemProtMD: Automated insertion of membrane protein structures into explicit lipid membranes. *Structure* 23:1350-1361.

Terpe, K. 2003. Overview of tag protein fusions: from molecular and biochemical fundamentals to commercial systems. *Applied Microbiology Biotechnology* 60:523-533.

Tsirigos, K. Hennerdal, A. Kall, L. Elofsson, A. 2013. A guideline to proteome-wide alpha-helical membrane protein topology predictions. *Proteomics* 12: 2282-2294.

Tsirigos KD. Govindarajan S., Bassot, C. Vastermark A. Lamb, J. Shu N. Elofsson A. 2018. Topology of membrane proteins – predictions, limitations and variations. *Curr. Opin. Structural Biology* 50: 9-17.

Tulumello DV. Deber CM. 2012. Efficiency of detergents at maintaining membrane protein structures in their biologically relevant forms. *Biochimica Biophysica Acta – Biomembranes* 1818: 1351-1358.

Turner, RJ. Weiner, JH. 1993. Evaluation of transmembrane helix prediction methods using the recently defined NMR structures of the coat proteins from bacteriophages M13 and Pf1. *Biochimica Biophysica Acta* 1202;161-168.

Viklund H, Elofsson A. Best alpha-helical transmembrane protein topology predictions are achieved using hidden Markov models and evolutionary information. *Protein Science* 2004;13:1908–1917.

Van Geest M. Lolkema JS. 2000. Membrane topology and insertion of membrane proteins: search for topogenic signals. *Microbiology Molecular Biology Reviews* 64: 13-33.

von Heijne, G 1992. Membrane protein structure prediction: hydrophobicity analysis and the positive-inside rule. *J. Molecular Biology* 225:487-494.

von Heijne, G. Rees, D. 2008. Membranes: reading between the lines. *Current Opinion Structural Biology*, 18;403-405.

von Heijne, G. 1999. A day in the life of Dr K. or How I learned to stop worrying and love lysozyme: a tragedy in six acts. *J. Molecular Biology* 293:367-379.

White SH. Wimley WC. 1999. Membrane protein folding and stability: physical principles. *Ann. Rev. Biophysics Biomolecular Structures* 28:319-365.

White, SH. 2015. Membrane protein topology: the messy process of guiding proteins into membranes. *eLife* 4:e12100.

Willegems, K. Efremov RG. 2018. Influence of lipid mimetics on gating of ryanodine receptor. *Structure* 26:1303-1313.

Winstone, TL. Duncalf, KA. Turner, RJ. 2002. Optimization of Expression and the Organic Extraction Purification of the Integral Membrane Protein EmrE. *Protein Expression Purification* 26: 111-121.

Yang Z, Wang C, Zhou Q, An J, Hildebrandt E, Aleksandrov LA, Kappes JC, DeLucas LJ, Riordan JR, Urbatsch IL, Hunt JF, Brouillette CG. 2014. Membrane protein stability can be compromised by detergent interactions with the extra membranous soluble domains. *Protein Science* 23:769-789.

Zhao G. London E. 2006. An amino acid "transmembrane tendency" scale that approaches the theoretical limit to accuracy for prediction of transmembrane helices: Relationship to biological hydrophobicity. *Protein Science* 15: 1987-2001.

INDEX

A

acid, 4, 8, 13, 20, 35, 36, 68, 82, 97, 100, 134, 143
acute myeloid leukemia, 73
adaptation, 15, 47, 64, 101
alkaline phosphatase, 5, 14, 85
amino acid, 3, 4, 7, 20, 35, 36, 44, 56, 59, 62, 69, 80, 81, 82, 95, 97, 102, 123, 131, 132, 134, 135, 136, 140, 146
angiogenesis, 18, 42, 47, 48, 70, 72, 73, 80
antibody, 29, 30, 51, 53, 59, 71
apoptosis, viii, 1, 9, 10, 14, 18, 19, 20, 22, 24, 37, 40, 42, 43, 45, 49, 54, 59, 62, 67, 68, 77
aqueous solutions, 144
aromatic rings, 125
atoms, ix, 92, 93, 95, 96, 101, 102, 104, 105, 107, 109, 110
autoantibodies, 23, 28, 59, 75
autoantigens, 73
autoimmune disease, 73
autoimmunity, 29, 89
autosomal dominant, 27
autosomal recessive, 27

B

bacteria, 104, 105
barriers, 47, 108, 110
bioavailability, 4, 10, 14, 64
biochemistry, vii, ix, 121, 122, 126, 128, 129, 137, 138
biological processes, viii, 91
biological responses, 9
biomarkers, 50, 55, 71, 78
biomolecules, 11, 132
blood, 5, 12, 14, 17, 18, 20, 21, 25, 29, 30, 41, 42, 47, 56, 71, 79
blood pressure, 20
blood vessels, 42, 47
body composition, 78
body mass index, 26, 31, 64
body size, 13
body weight, 76
bone marrow, 57, 86
breast cancer, 46, 47, 54, 59, 64, 83, 84
budding, ix, 121, 122

C

cancer, 5, 7, 9, 11, 12, 37, 38, 39, 40, 41, 42, 44, 45, 46, 47, 48, 49, 50, 51, 52, 53, 55, 56, 57, 58, 59, 60, 64, 65, 66, 67, 69, 71, 74, 75, 77, 78, 79, 80, 82, 83, 84, 88, 89, 104
cancer cells, 37, 39, 40, 41, 42, 43, 44, 45, 48, 49, 53, 59, 64, 65, 71, 88, 104
cancer progression, 60
cancer therapy, 75, 89
carbohydrate, 11, 30, 31, 61, 78, 84
carbohydrate metabolism, 11
carcinoma, 15, 78, 81, 85, 89
cell death, 37, 43, 50, 69
cell division, 46
cell invasiveness, 47
cell line, 18, 19, 21, 48, 51, 54, 57, 64, 88
cell membranes, 11, 37, 94
cell metabolism, 44, 50, 68
cell signaling, 80, 110
cell surface, 6, 14, 24, 28, 35, 52, 61, 82
cell surface proteoglycans, 82
chemical, 32, 93, 95, 105, 133
chemical bonds, 105
chemokine receptor, 102
choriocarcinoma, 18, 61, 79
chromatography, 127, 128
circulation, 4, 11, 18, 33, 43, 46, 51
colon, 2, 14, 41, 51, 52, 64, 81, 89
colorectal cancer, 51, 60, 66, 84, 85
copolymer, 126, 143
correlation, 30, 33, 75, 93, 97
correlation function, 97
crystal structure, 85, 103, 104, 109, 110
crystallization, 109, 129, 141, 142, 143
cytoplasm, 19, 20, 100, 123
cytostatic drugs, 53

D

database, 116, 137, 143
defects, 24, 31, 56, 93
degradation, 10, 18, 23, 48, 52, 123
diabetes, 2, 14, 15, 17, 18, 22, 23, 24, 25, 26, 28, 30, 32, 34, 35, 36, 55, 56, 57, 58, 59, 60, 61, 62, 63, 64, 65, 66, 67, 68, 69, 70, 72, 73, 74, 75, 76, 77, 78, 80, 81, 82, 83, 84, 86, 87, 88
diabetic ketoacidosis, 27
diabetic neuropathy, 34
diabetic patients, 32, 76
dielectric constant, 129, 132, 140
diffusion, 93, 94, 106, 108, 113, 131
diseases, 15, 37, 59
distribution, 8, 35, 61, 85, 99
DNA, 42, 43, 69, 118, 143
drug discovery, 64
drug targets, viii, 91
drugs, 34, 110

E

electron microscopy, 126, 139, 144
endothelial cells, 18, 47, 73
endothelium, 17, 18, 35, 61, 64
energy, viii, 1, 23, 34, 44, 50, 91, 92, 96, 99, 108, 110, 131
environment, 41, 45, 47, 52, 61, 83, 93, 102, 105, 115, 116, 125, 127, 128, 134, 135, 137, 145
epithelial cells, 51, 65
epithelial ovarian cancer, 59, 83
equilibrium, 12, 13, 50, 93, 97, 128
eukaryotic, 104, 108, 137
evidence, 15, 24, 32, 57, 61, 66
extracellular matrix, 11, 41, 47, 48, 50, 52, 56

Index

F

fetal growth, 61, 67, 70, 72, 83, 84
fibroblasts, 16, 17, 31, 81
force fields, 92, 109
formation, 14, 43, 47, 79
fragments, 43, 52, 62
freedom, 95, 96, 131
fusion, 16, 42, 99, 114, 118, 126, 137, 144

G

gene expression, 12, 57, 58, 67, 83
genes, 3, 22, 40, 47, 53, 56, 59, 66, 67, 73, 80, 82, 89, 123
genetic disease, 104
genome, viii, 40, 45, 69, 73, 91, 92
gestation, 5, 16, 18, 20, 34, 81
gestational age, 21, 22, 69
gestational diabetes, 18, 55, 56, 59, 60, 62, 66, 74, 76, 82
glucose, 11, 20, 23, 24, 27, 29, 31, 34, 35, 36, 44, 54, 62, 79, 82, 88
glucose tolerance, 23, 88
glucose tolerance test, 88
glycans, 6, 18, 19, 22, 35, 36, 53, 85
glycerol, 103, 133, 139
glycine, 101, 102, 114
glycogen, 8, 61, 83
glycol, 126, 142
glycolysis, 44
glycoproteins, 54, 81
glycosaminoglycans, 11
glycosylation, 6, 8, 10, 14, 18, 50, 52, 60, 63, 64, 65, 68, 69, 72, 78, 81, 123
growth, vii, 1, 2, 3, 5, 6, 7, 8, 9, 10, 11, 12, 13, 14, 15, 16, 18, 19, 21, 24, 27, 31, 36, 37, 38, 39, 42, 45, 46, 47, 48, 49, 53, 54, 55, 56, 57, 58, 59, 60, 61, 62, 63, 64, 65, 66, 67, 68, 69, 70, 71, 72, 73, 74, 75, 76, 77, 78, 79, 80, 81, 82, 83, 84, 85, 86, 87, 88, 89, 110, 137
growth factor, vii, 1, 2, 6, 11, 36, 45, 55, 57, 58, 60, 61, 62, 63, 64, 65, 66, 67, 68, 69, 70, 71, 72, 73, 74, 75, 76, 77, 78, 79, 80, 81, 82, 83, 84, 85, 86, 87, 88, 89
growth hormone, 3, 12, 15, 54, 56, 57

H

human, 15, 16, 18, 19, 20, 48, 49, 55, 57, 58, 59, 61, 62, 64, 65, 66, 67, 68, 69, 71, 72, 73, 74, 76, 77, 78, 79, 81, 82, 83, 84, 85, 88, 89, 114, 118
human immunodeficiency virus, 114
hybrid, 5, 9, 13, 57, 63, 68, 85, 86, 118
hydrogen, 102, 106, 113, 128
hydrogen bonds, 106, 128
hydrophilicity, 135
hydrophobicity, 97, 134, 142, 146
hydroxyl, 102
hyperandrogenism, 27, 29
hyperglycaemia, 24, 27, 29, 54
hyperinsulinemia, 27, 31, 54, 63
hypermethylation, 21, 75
hyperplasia, 27, 35, 80
hypertension, 13, 31, 62, 86
hypertrophy, 80
hypoglycemia, 59, 75
hypothesis, 26, 31, 32, 71, 74
hypoxia, 21, 43, 47, 63, 72, 89

I

immune response, 45
immune system, 44, 45, 49
immunoglobulins, 29
immunosuppression, 29
imprinting, 3, 46, 59, 60, 71, 78
improvements, 135, 136, 140
in vitro, 19, 24, 46, 59, 76, 79, 128, 134

in vivo, 63, 65, 79, 128, 131, 133, 134, 137, 144
inflammation, 23, 34, 44, 45, 65, 80
influenza, 102, 112, 116, 118
infrared spectroscopy, 99
inhibition, 8, 15, 18, 37, 40, 69, 72, 80, 85, 89, 131
insertion, 124, 135, 136, 142, 145
insulin, vii, 1, 2, 3, 5, 6, 8, 9, 11, 12, 13, 15, 18, 20, 23, 24, 26, 27, 28, 29, 30, 31, 33, 34, 35, 36, 47, 54, 55, 56, 57, 58, 59, 60, 61, 62, 63, 64, 65, 66, 67, 68, 69, 70, 71, 72, 73, 74, 75, 76, 77, 78, 79, 80, 81, 82, 83, 84, 85, 86, 87, 88, 89
insulin resistance, 23, 24, 26, 27, 28, 29, 30, 31, 34, 36, 37, 60, 63, 64, 66, 70, 74, 75, 80, 82, 83, 89
insulin sensitivity, 15, 23, 24, 25, 26, 30, 54, 58, 63, 67, 76, 77, 85
insulin signaling, 60, 61
insulin–like growth factors, 2, 60, 61, 64, 67, 70, 74, 78, 84, 88
interface, 58, 66, 69, 99, 114
ion channels, viii, 91, 98, 101, 102, 115, 116, 133, 136
ion transport, 114
ion/cation concentrations, 92
ion-exchange, 127
ions, 98, 107, 109, 110, 115, 125, 127, 129, 132
issues, 64, 124, 126, 127, 128, 129, 131, 132, 134, 141

K

karyotype, 37, 42
keratinocytes, 31
ketoacidosis, 27
kinase activity, 8, 10
kinetics, 96, 106, 131

L

ligand, 5, 6, 8, 9, 39, 50, 53, 57, 71, 75, 109, 110, 119, 125, 128, 129, 131, 132, 133, 143, 145
lipid, viii, 8, 11, 31, 35, 61, 68, 79, 84, 92, 93, 100, 101, 105, 106, 109, 110, 112, 113, 115, 116, 117, 119, 120, 124, 126, 127, 128, 130, 131, 132, 133, 136, 137, 138, 139, 140, 142, 143, 145, 146
lipid metabolism, 11, 84
liver, 3, 8, 11, 21, 23, 27, 37, 41
liver cancer, 37
liver enzymes, 21

M

macrosomia, 21, 34, 35, 36, 70
majority, 7, 11, 42, 51, 138, 140
melanoma, 40
mellitus, 14, 18, 55, 56, 57, 58, 59, 60, 61, 62, 63, 65, 66, 67, 68, 70, 72, 73, 74, 76, 77, 80, 81, 82, 83, 84, 86, 87
melting temperature, 97
membrane curvature, 92, 94, 126
membrane potentials, viii, 91, 92
membrane proteins, v, vii, viii, 1, 18, 19, 22, 52, 81, 91, 92, 94, 99, 100, 101, 102, 103, 104, 105, 107, 108, 109, 110, 111, 118, 119, 121, 122, 132, 141, 142, 143, 144, 145
membrane receptors, 2, 11, 12, 15, 54
membranes, 26, 66, 71, 94, 98, 99, 105, 108, 111, 114, 131, 140, 143, 145, 146
mesenchymal stem cells, 21, 88
messenger ribonucleic acid, 84
metabolic disorder, 15
metabolic pathways, vii, 1
metabolic responses, 11
metabolic syndrome, 35, 58

metabolism, viii, 1, 3, 11, 20, 24, 31, 35, 36, 47, 54, 60, 79
metastasis, 41, 50, 54, 55, 75, 82, 84
migration, 10, 15, 17, 19, 20, 48, 54, 61, 68, 73, 76, 125, 129, 144
model system, 54, 130, 134
models, 32, 94, 95, 96, 98, 100, 101, 102, 103, 109, 110, 134, 145
modifications, 4, 5, 7, 10, 14, 19, 51, 87, 123
molecular biology, 127
molecular dynamics, viii, 92, 100, 102, 104, 105, 111, 113, 114, 115, 116, 117, 119, 120, 140
molecular structure, viii, 92, 143
molecular weight, viii, 91, 125, 129, 132
molecules, 4, 5, 16, 17, 33, 39, 41, 44, 45, 46, 50, 53, 92, 95, 97, 99, 102, 108, 120, 129, 138
monolayer, 100
mortality, 20, 27, 37, 42
motif, 73, 102, 103, 104, 116

N

nucleic acid, 54, 132
nucleic acid synthesis, 54
nucleotides, 44
nucleus, 9, 11, 53, 83

O

obesity, 13, 15, 23, 29, 32, 34, 58, 67, 68, 78, 79
oxidation, vii, viii, 1, 14, 15, 21, 32, 52, 65
oxidative stress, 15, 20, 22, 34, 50, 52, 57, 127
oxygen, 15, 16, 20, 21, 40, 42, 44, 47, 50, 88

P

pathways, 8, 11, 12, 14, 15, 20, 22, 24, 35, 47, 48, 50, 68, 74, 80, 103, 109, 136
permeation, 98, 103, 108, 114, 117
phenotype, 38, 41, 45, 50, 53, 55, 61, 84, 134
phosphate, 10, 13, 62, 72, 74, 77, 78, 79, 84
phosphatidylcholine, 99
phosphorylation, 6, 7, 8, 14, 29, 44, 65, 68, 70, 75, 76, 83, 88
physical phenomena, 96
physical properties, 96, 139
physiology, 11, 57, 58, 85, 89
placenta, 2, 3, 13, 15, 16, 17, 18, 19, 20, 22, 31, 34, 35, 36, 55, 58, 60, 61, 64, 66, 67, 68, 70, 71, 72, 76, 79, 81, 85, 86, 89
placental abruption, 21
placental structures, 16
plasma membrane, 6, 7, 8, 18, 86, 110, 112, 143
polar, 95, 108, 125, 126, 133, 135
polypeptide, 97, 104, 129, 137, 139
pregnancy, 5, 14, 15, 17, 18, 20, 34, 35, 58, 67, 71, 76, 85, 88
principal component analysis, 108
proliferation, viii, 1, 3, 6, 7, 9, 15, 16, 18, 19, 20, 21, 31, 35, 36, 37, 38, 39, 40, 47, 54, 61, 71, 76, 86, 88, 89
protein structure, viii, 91, 93, 102, 108, 110, 111, 138, 141, 142, 145, 146
protein synthesis, 49, 89
protein-protein interactions, 138
proteins, vii, viii, 1, 2, 3, 4, 6, 8, 11, 12, 15, 18, 19, 20, 22, 28, 36, 40, 43, 45, 46, 51, 52, 55, 56, 60, 63, 70, 78, 81, 88, 89, 91, 92, 94, 97, 98, 99, 100, 101, 102, 103, 104, 105, 107, 108, 109, 110, 111, 112, 118, 119, 120, 122, 123, 125, 127, 129, 132, 136, 137, 138, 140, 141, 142, 143, 144, 145, 146

purification, vii, ix, 121, 127, 131, 139, 141, 143

R

receptor, 2, 5, 6, 7, 8, 9, 10, 14, 22, 24, 26, 27, 28, 29, 32, 33, 34, 35, 46, 47, 48, 50, 52, 54, 56, 57, 58, 59, 60, 61, 62, 63, 64, 65, 66, 67, 68, 69, 71, 72, 73, 74, 75, 76, 77, 78, 79, 80, 82, 83, 84, 85, 86, 87, 88, 89, 112, 131, 146
recognition, viii, 3, 91, 113, 136
recommendations, iv
researchers, 33, 124, 134, 139
residues, 6, 8, 13, 22, 104, 134
resistance, 9, 14, 22, 25, 27, 28, 29, 31, 32, 33, 37, 53, 61, 68, 77, 88, 89, 104, 118, 128, 144
resolution, vii, ix, 104, 110, 121
respiratory syncytial virus, 116
response, 6, 21, 23, 29, 47, 51, 53, 56, 69

S

secretion, 23, 24, 25, 35, 67, 80, 85
selectivity, viii, 91, 107, 109, 114, 136
semi-explicit methods, 92
sensitivity, 23, 25, 26, 30, 33, 45, 49, 65, 71, 115
sensor, 36, 43, 70, 114, 115, 119
serum, 14, 29, 47, 75, 79, 86
signal transduction, 8, 29, 50, 58, 62, 69, 99, 100
signaling pathway, 49, 60, 88, 89
signalling, 6, 7, 8, 9, 10, 11, 12, 14, 15, 16, 19, 20, 21, 22, 23, 24, 32, 34, 35, 36, 38, 39, 40, 43, 45, 48, 49, 50, 53, 55, 60, 64, 66, 71, 72
signals, vii, 1, 6, 7, 9, 19, 38, 39, 40, 41, 43, 45, 48, 50, 54, 145

simulations, vii, viii, 92, 93, 94, 96, 97, 98, 99, 100, 101, 102, 103, 104, 105, 108, 109, 110, 111, 112, 113, 114, 115, 116, 117, 118, 119, 140
skeletal muscle, 11, 26, 30, 63, 71, 77
sodium dodecyl sulfate, 129
sodium dodecyl sulfate (SDS), 129
stability, 42, 47, 104, 126, 127, 128, 129, 146
stress, 22, 24, 50, 52, 63, 80, 89, 110, 123
structural changes, 52, 98, 108
structural modifications, vii, 1
structure, viii, ix, 5, 16, 35, 40, 50, 51, 52, 60, 68, 78, 85, 86, 92, 94, 96, 97, 101, 104, 105, 113, 114, 116, 118, 121, 122, 127, 129, 131, 134, 137, 138, 139, 140, 144
surfactants, 124, 126, 144
surveillance, 44, 45, 81
survival, 7, 9, 12, 20, 28, 36, 43, 47, 54, 74, 83
synaptic transmission, 108
syndrome, 21, 27, 58, 63, 70, 75, 87
synthesis, 8, 11, 27, 44, 68, 83

T

therapeutic approaches, 15
therapeutic targets, 14, 15, 67
therapy, 30, 36, 37, 53, 55
thrombocytopenia, 20
thrombocytosis, 77
thrombosis, 32, 87
thyroid, 13, 37, 87
thyroid cancer, 13, 37, 87
tissue, vii, viii, 2, 16, 18, 26, 32, 34, 35, 37, 39, 41, 44, 45, 46, 47, 50, 52, 69, 79, 81
tissue homeostasis, 37
topology, viii, 91, 97, 123, 130, 133, 136, 137, 140, 141, 143, 145, 146
transformation, 10, 37, 38, 43, 50, 53

translocation, 7, 20, 24, 25, 106, 143
transmembrane, viii, 5, 7, 10, 91, 92, 98, 101, 104, 114, 115, 116, 123, 129, 133, 135, 137, 139, 143, 145, 146
transmembrane region, 135
transport, viii, 4, 6, 15, 35, 36, 56, 58, 62, 70, 82, 91, 98, 102, 108, 109, 118, 119, 123
treatment, 14, 15, 29, 30, 35, 57, 82, 87, 105
tumor, 59, 60, 63, 69, 71, 75, 78, 87
tumor cells, 60
tumor development, 78
tumor invasion, 75
tumorigenesis, 70
type 1 diabetes, 23, 35

type 2 diabetes, 25, 26, 28, 34, 56, 59, 61, 67, 74, 83, 86, 88
tyrosine, 5, 6, 7, 8, 10, 14, 23, 60, 68, 69, 72

W

water, 93, 95, 97, 99, 102, 103, 106, 107, 109, 117, 127, 129, 139, 144
workers, 9, 19, 28, 35, 53
worldwide, 20, 23, 34, 51
wound healing, 14, 33, 43, 45